Robert Dundas Thomson

Experimental researches on the food of animals and the fattening of

cattle

With remarks on the food of man

Robert Dundas Thomson

Experimental researches on the food of animals and the fattening of cattle
With remarks on the food of man

ISBN/EAN: 9783337201302

Printed in Europe, USA, Canada, Australia, Japan

Cover: Foto ©berggeist007 / pixelio.de

More available books at **www.hansebooks.com**

TO

DR. THOMAS THOMSON

AND

BARON LIEBIG,

TO WHOM THE AUTHOR OWES HIS ACQUAINTANCE

WITH THE SCIENCE OF CHEMISTRY,

This Contribution

TOWARDS THE DEVELOPMENT OF THE SUBJECT OF THE

GROWTH OF ANIMALS

IS

AFFECTIONATELY INSCRIBED.

PREFACE.

THE present Work is based on an extensive series of experiments which were made at the instance of the Government. The original object of that inquiry was to determine the relative influence of barley and malt in feeding cattle ; but as the opportunity seemed a favorable one for investigating some scientific problems of great importance to physiology, and of extreme value in the physical management of man and animals, advantage was taken of it, by permission, to extend the experiments so as to include these objects.

It is well known to those who have been in the habit of late years of following the researches which have been undertaken to elucidate the nature of the growth of animals, that it is now generally agreed that the muscular part of animals is derived from the fibrinous or nitrogenous ingredients of the food, while the source of animal fat has been disputed. The present experiments seem to demonstrate that the fat of animals cannot be produced from the oil of the food, but must be evolved from the calorifient, or heat-forming portion of the aliment, essentially assisted by its nitrogenous materials. By following out this principle, the author has been enabled to detect an important relation subsisting between the nutritive and calorifient portion of the food, upon the determination of which, for the various conditions of animals, he considers the laws of animal dieting depend. He has endeavored to apply this law to various articles of human food ; and he trusts that the basis has been laid for future researches, which may be directed to administer to the health and comfort of mankind, and of domesticated animals. In conducting the experiments upon cattle, the author found not only his habitual acquaintance with animals, but also his medical knowledge in continual requisition in consequence of the tendency of the varied conditions of the animal system, from the sudden and frequent changes of diet, to induce symptoms of disease. These were carefully watched, and overcome by such precautions as clearly follow from a due consideration of the principles announced in this work. It was on this account, and to enable the agriculturist to appreciate the advantage which he would derive from physiological and chemical knowledge, rather than to give anatomical instruction to the professional man, that the introductory chapters were written. In a work professing to be the result of entirely original experiments, and where such a mass of figures exist, errors must unavoidably have been overlooked, even although great care has been taken to diminish their number. The author, however, trusts that none will be detected which can materially interfere with the principles deduced from the researches.

CONTENTS.

CHAPTER I.

CHAPTER II.

CHAPTER III.

CHAPTER VII.

CHAPTER VIII.

CHAPTER IX.

APPENDIX.

RESEARCHES

ON

THE FOOD OF ANIMALS,

&c. &c.

CHAPTER I.

INTRODUCTION.—DIFFERENT EXPLANATIONS OF DIGESTION.—IMPORTANCE
OF RESEARCHES TO DISCOVER ITS TRUE NATURE.—SIMPLICITY OF LIV-
ING, AND NOT THE SAVAGE LIFE, CONDUCIVE TO HEALTH.

It is a remark no less old than true, That we are
often less acquainted with the nature of facts of every-
day occurrence, than with those of a rarer description.
This may proceed from one of two causes ; either from
the phenomena constantly under our notice being neglected, in consequence of our familiarity with them, or
from the complexity of their nature, and the intricate
purposes which they ultimately subserve. Some physiologists, who have endeavored to explain the nature
of the process of digestion, would ascribe our ignorance
of that important function to the former of these causes ;
since they refer the preparation of the food in the stomach
for the purpose of nourishing the body to the presence
in that organ of an acid, which, according to them, simply dissolves the food, and enables it to enter as a constituent of the circulating fluids of the animal system.
The acid which effects this important object is the hy-

2

drochloric acid; which they consider to have been satisfactorily proved to be present during the period when
food exists in the stomach, and they conceive that they
can imitate the process of animal digestion in glass, or
other vessels out of the body, simply by exposing animal and vegetable food to the influence of dilute acids.
Another class of individuals, who have studied the interesting changes which the food undergoes in the
stomach and intestines, conceive that we are still unacquainted with the true nature of this process, and are
inclined to the opinion that the reason why we are not
sufficiently conversant with the phenomena of digestion,
depends more on their intricacy and obscurity than upon
a deficiency of research and observation ; and that while
we possess some facts which seem to indicate the direction in which we are to search for a solution of the
difficulties of the subject, we are still at a great distance
from the elucidation of the precise manner in which
animals digest their food.

There cannot be a doubt that if we understood the
nature of the process by which the food which we
swallow is converted into living flesh, important results
would follow in reference to the preservation of the
health of animals, and the treatment of diseases. If we
were properly acquainted with every transformation
through which the constituents of the food pass after it
has been masticated, until it is finally removed from the
system, it is clear that, in cases where the stomach is
unable to perform its accustomed functions, the assistance of art might be called in to minister to digestion.
Even in the present state of our knowledge, civilized
nations cook their food, or, in other words, endeavor to
imitate the primary stage of digestion, while the savage

in his wild, untutored state, being in a condition akin
to that of the beasts of the forest, scarcely stands in
need of the assistance of art, and devours his prey with
less of enjoyment than of necessity.

It has been a favorite speculation with some philosophers, that as beasts thrive best in the forest, so man is
most healthy in the savage state ; that when accustomed
to brave the severity of the winter's cold and summer's
heat, to contend with the snow and the thunder storm
without the protection of clothing, or pampering food,
he is armed, like the Spartan of old, with a shield
against the disease and early death so prevalent among
the members of refined societies ; that the catalogue of
maladies existing among a primitive people is exceed-
ingly limited, and that it augments in volume precisely
in proportion to the encroachments of civilization, and
to the departure from those simple laws by which nature, in her unsophisticated state, is uniformly guided.
So far has this view been carried by some advocates,
that it was the opinion of Plato, that after certain medicines were introduced by Podalirius and Machaon at
the siege of Troy, different diseases, which these medicines produced, became prevalent. It can scarcely be
denied, that while these opinions are founded in truth,
they have been greatly exaggerated, and made to tell in
the wrong direction. It is quite true that simplicity in
diet is better fitted to perpetuate health than stimulating
and unnatural food ; but it is not necessary that, in order to acquire health, man should return to the actual
condition of the savage ; nor is it incumbent that, although our domestic animals are seen to thrive well in
their primitive forests, they should be cast loose under
literally the same circumstances. In other words, it

does not follow, because savage man and animals are healthy, that civilized man and his attendant animals should be diseased. A little reflection will show, that a greater amount of knowledge is required to manage animals which are subjected to artificial restraints than in their original condition; for while man in a social state undergoes more mental and physical fatigue than in a state of mere nature, so his attendant animals being placed under certain restrictions, foreign as it were to their primitive condition, it is necessary for those who direct their attention to the management of the physical nature of both man and animals, to possess such an acquaintance with their construction and requirements, as to be able to lay down regulations for retaining them in a healthy and natural condition of body, and to prevent cattle, more especially, from acquiring that unwholesome fat condition which, from want of due attention to the nature of the animal's system, has assumed almost the aspect of a permanent fallacy.

To render the doctrines to be laid down in the subsequent part of this work more intelligible, it will be proper to describe briefly the organs of digestion in man and cattle, and to notice the opinions entertained respecting the nature of digestion. In accomplishing this, it will be necessary to distinguish between what is known and what is assumed.

CHAPTER II

HUNGER AND THIRST ARE LAWS OF NATURE.—ANECDOTE.—MASTICATION OR CHEWING NECESSARY AS A PREPARATION FOR DIGESTION.—IMPORTANCE OF THE FINE DIVISION OF FOOD FOR THE PRODUCTION OF MILK IN COWS.—EXPERIMENT ILLUSTRATIVE OF THIS POSITION.—ALCOHOL NOT NECESSARY IN HUMAN AND ANIMAL DIET.—ANECDOTE OF A FOREIGNER.—DEFINITION OF DIGESTION.

HUNGER and thirst are the preliminary steps to digestion; they constitute a law implanted in the animal economy for the purpose of inducing the living being to take such nourishment as is required to sustain that waste of the system which animated nature is continually undergoing. If the dictates of the sensation of hunger and thirst are rationally obeyed, satisfaction and healthy digestion are the result; but if, on the contrary, these important sensations are neglected, weakness and disease must necessarily ensue. Appetite, or, in its more advanced stage, hunger, teaches animals to seek for solid food, and thirst suggests the propriety of rendering the solid mass more pulpy and dilute by the employment of drink. Experience and reason, both in man and brutes, must in some measure direct the selection of the proper objects to be employed for these purposes. I was some years ago consulted by a worthy individual with regard to the propriety of fasting as a religious observance. I told him that the sensation of hunger and thirst constituted a most important law in the animal economy, destined by the Creator for the

most beneficent purposes; that it ought to be obeyed
as a matter of duty, and that if infringed, some preju-
dicial result would necessarily ensue; because it is no
argument in favor of any such experiment upon human
life that existence does not terminate upon its adoption,
or that the symptoms of some frightful disease are not
instantly ushered in. The seeds of future mischief
may be sown by one experiment, and may only lie dor-
mant until a second or succeeding infringement shall
cause them to spring forth into living activity. In the
course of the extensive series of experiments upon
cows afterwards to be detailed, it was found that, when
they were not supplied with sufficient food during one
day the product of milk was a day or two in reaching
its former average; thus demonstrating that the animal
had been weakened by the abstinence, inasmuch as it
took a longer period to reach its ordinary condition than
was required to reduce it. The milk, in such an experiment, corresponds with the muscle and fatty portions of the body of animals which do not supply milk;
hence abstinence in all animals must be followed by a
diminution of the weight of the body. It has been
well remarked by Liebig, that "in the process of starvation it is not only the fat which disappears, but also
by degrees all such of the solids as are capable of being dissolved. In the wasted bodies of those who have
suffered starvation, the muscles are shrunk, and un-
naturally soft, and have lost their contractility : all these
parts of the body which were capable of entering into
the state of motion have served to protect the remainder of the frame from the destructive influence of the
atmosphere." (*Liebig*, p. 26.) There is no difference
in this respect between one set of animals and another.

Civilized and savage men, wild and domestic animals, must all be classed under the same category.

In the human species a morsel of food is grasped by the front teeth of both jaws, which are each supplied with sixteen teeth, making thirty-two in all. In those animals which chew the cud, as they have only one row of teeth the food is less firmly grasped by the jaws, and there is, therefore, a greater necessity that it should be of a soft and pliable nature. By the assistance of the lips, jaws, tongue, and auxiliary muscles, the food is conveyed into the cavity of the mouth, and by the aid of the tongue and lateral motion of the mouth it is placed between the opposing jaws, where it is masticated or ground to a proper consistence. But the action of the jaws in grinding the morsel introduced between them at the same time elicits the compressing power of the muscles of the cheek upon the parotid gland, which is situated in man in front of the ear, and expels its secreted fluid, the saliva, into the mouth, to assist in comminuting the nutritive matter. Besides this mechanical action, there is, however, a nervous sympathy called into operation. The masticated matter acts upon the tongue and adjacent parts, inducing a sympathy with the glands placed under the tongue, and causes them to pour out their copious contents. The object of mastication or chewing is, therefore, to reduce the food to such a consistence as shall fit it for its reception and proper digestion in the stomach. This is well illustrated in the instance of animals which are not supplied with teeth.

The common fowl, for example, is destitute of these grinding apparatus; but it has a muscular mechanism termed the gizzard, which powerfully compresses the introduced food, and by means of pebbles and stones, which are a necessary article of food with the class of animals referred to, an artificial substitute for the teeth is provided. In graminivorous animals, we shall presently find that a substitute for the second row of teeth is provided in the operation of rumination, or chewing the cud. From attention to these facts, therefore, we are taught that the preparatory step of digestion consists in the fine division of solid food by means of the apparatus set apart in the mouth for this purpose, and its mixture with a certain amount of fluid saliva to render it more dilute.

The importance of the proper grinding of the food, and of rendering it as soluble as possible, can be well appreciated by such individuals as have been the subjects of indigestion, from the eructation of morsels of food, of gases, and of acid liquors. It is scarcely necessary to remark, that similar rules are applicable to the inferior animals, and more particularly in the state of confinement to which most of them are more or less subjected when they are made to minister to the wants of the human species. The following comparative table exhibits this fact in a sufficiently striking manner. Two cows were fed on entire barley and malt, steeped in hot water; they were then fed on crushed barley and malt, prepared in the same manner. The influence of the finer division of the grain in augmenting the product of milk places the importance of this position beyond all cavil :—

	BROWN COW. Milk in Periods of 5 Days.	WHITE COW. Milk in Periods of 5 Days.
Entire barley and grass, - -	$\{$ $111\frac{1}{2}$ lbs. $97\frac{1}{3}$	106 lbs. 94
Entire malt and grass, - -	$\{$ 96 95	98 104
Crushed barley, grass and hay,	$\{$ $115\frac{1}{4}$ 105 110	$109\frac{1}{2}$ $109\frac{1}{3}$ 110
Crushed malt and hay, - -	$\{$ 97 96 98	$106\frac{1}{2}$ $107\frac{1}{2}$ $111\frac{1}{2}$

An inspection of this table shows, that with the entire barley the milk diminished during the second five days of the experiment, while with the crushed barley the milk had a tendency to increase during each succeeding period. In all such experiments there are continually occurring irregularities, of which we have no means of precisely appreciating the causes. These proceed often from atmospherical influences, as temperature, and frequently from the condition of the animal. We are, therefore, taking a legitimate view of an experiment, when we direct our views to the tendency to improvement or deterioration in the course of the trial, rather than to the actual numbers obtained. In the preceding table, the tendency to an increase of product is decidedly in favor of the finely divided grain. There are some anomalies, more particularly with reference to the brown cow, which was rather a fiery animal, and probably placed in peculiar physical conditions, as will subsequently be explained.

The nature of the saliva, which is a fluid of the simplest constitution, as it contains $99\frac{1}{2}$ per cent. of water, directs our attention to the nature of the fluid to be used

in quenching thirst.   It has become customary in towns
to stimulate the systems of cattle, more especially of
cows, after the fashion of human beings, by the use of
alcoholic fluids, such as pot ale, under the idea of increasing the amount of milk.   Now as the stimulating
portion of this pot ale is alcohol, and contains no curd,
or, if so, but an insignificant portion, it is evident that
no increase of the nutritive constituents of the milk is
thereby obtained.   It is an idea, too prevalent with
nurses, that fermented liquors increase the quantity of
milk ; but I am sure all intelligent physicians will agree
with me that this view should not be encouraged, either
as improving the quality of the milk, or as benefiting
the infants supported on such food.   Even for adults a
similar advice may not be inappropriate.   A foreigner,
who had a high opinion of English philosophy, was in-
vited to a party consisting of men of science.   After a
plenteous dinner the table was cleared, and the bottles
were placed on the table.   Having partaken of two or
three glasses of wine, and being still pressed to drink,
he seriously assured the company that his thirst was
quenched.   The philosophers, however, continued to
urge him to follow their example, and drink, even although he were not thirsty ; upon which the foreigner
rang the bell, and insisted on having another course
brought up, declaring, that they ought to eat as much
against reason, as he to drink.   The only advantage
gained can merely be by stimulating the system, or in
supplying a bad form of heat-producing food in a liquid
form.   There is no evidence that alcohol can supply
any of the constituents of the milk or body.   If the
milk augments under its action, a position requiring to
be proved, it must be in regard to the aqueous ingre-

dient, and not by an increase of any of the solid constituents ; a consequence, therefore, which would be more
satisfactorily acquired by the addition of water to the
milk after it has been drawn from the animal.

The saliva would appear to constitute the type of
what the drink of man and animals should be. The
artificial beverages so much employed by them in a
state of confinement seem to be unnecessary, if not
hurtful. By the use of fluids as nearly allied to the
nature of saliva as possible, we shall, as far as we can
judge, be following the simple rules of nature. The
operation of mastication, or chewing, is a voluntary act ;
but the next step, or that of deglutition, or swallowing,
is of a different character. So soon as the food is sufficiently reduced to a pulpy state, the natural impulse
appears to be to carry it, by the assistance of the
tongue, to the back part of the mouth. This is all the
voluntary exertion required on the part of the individual.
The instant that it touches certain nerves which guard
the throat, they are excited, and cause the muscles to
grasp the morsel and carry it into the gullet, by which
it is conveyed, without any peculiar sensation, in the
healthy condition of animals, and without any exercise
of voluntary motion, into the stomach, the primary organ of digestion.

Much ambiguity has occurred in physiological wri-
tings respecting the nature of digestion, perhaps as much
from the absence of a proper definition of the term as
from any other cause. Some writers appear to consider
the disappearance of the masticated food from the stomach as a proof of the completion of the process of digestion, while others view digestion as the formation of a
pulpy mass in that organ. Physiologists generally de-

scribe the pulpy mass in the stomach under the name of chyme, and that in the smaller intestines as chyle ; but as these terms are in some measure artificial, and scarcely admissible in the case of graminivorous animals, in the subsequent description of what is known respecting the changes which the food undergoes in the intestines, these terms will be omitted. By digestion I understand the conversion of food into blood. A consideration of this subject will lead us to notice the principal organs of digestion in man and animals, as well as the primary steps of digestion in the stomach and intestines, with the secondary stage of digestion in the passage of the food to the blood-vessels, and the alteration which it there undergoes.

CHAPTER III.

Human Organs of Digestion.—The organs of primary digestion in man are all situated in the lower
division of the trunk of the body, usually termed the
abdomen or belly, (*Fig.* 1.) They consist of the
stomach, which may be viewed as an expansion of the
gullet, or meat-pipe. Its form has been compared to
that of a bagpipe. It lies principally on the left side,
under the edge of the ribs ; but it extends towards the
middle of the body, and more particularly after a meal
its expansion can be detected. The upper border of
the stomach is curved ; the hollow of the curve extend-
ing downwards, and forming what is designated the
small curvature or arch of the stomach. The lower
border of this organ also constitutes an arch, termed the
greater curvature. The passage into the stomach from
the gullet, and the exit-valve or intestinal or lower ex-
tremity of the stomach are thus nearly on a level, so
that this organ may be said to be directed across the

3

body.   The lower opening of the stomach (pyloric ori-
fice) is contracted, being supplied with a circular band

Fig. 1.

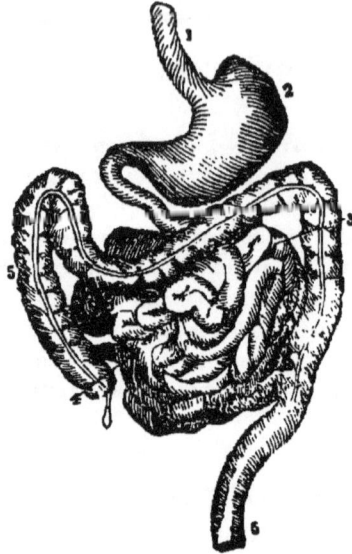

HUMAN STOMACH AND INTESTINES, (Grant.)

1. Œsophagus, or meat-pipe.
2. Stomach.
3. Small intestines.
4. Termination of the small intestines in the colon
5. Great arch of the colon.
6. Straight gut, or rectum.

of muscular fibres, which constitutes a kind of valve in
order to prevent food from returning into this organ.
This point forms also the connection with the intestines,
from whence they extend in the form of a long tube,
five or six times the length of the body, and occupy the
lower part of the abdomen.   The intestines are usually
divided into the small and large intestines.   The former
‘ are estimated to be in length twenty-six feet, or from

four to five times the length of the body ; and the great intestines one length of the body, or about six feet."— (*Bell.*) But it is rather remarkable that we have no precise statistical data in reference to the proportion between the height of the body and the length of the intestinal canal. In the figure the small intestines occupy the middle space, and are surrounded on three sides by the large intestines. The colon, which commences on the right side of the body, passes upwards and across to the left side, in the form of a great arch ; then downwards, until it terminates in the rectum, or straight gut. The upper portion of the small intestines is termed duodenum, from its being twelve finger-breadths in length. It crosses over to the right side of the spine, and descends to the kidney, from which it crosses over to the left side of the spine. This is the largest of the small intestines, and it generally contains digested matter. The next portion of the small viscera, or two-fifths of what remains, is termed the *jejunum*, or empty intestine, because it is generally void of contents. The lower portion of the small intestines is termed ilium, and resembles the *empty* intestine. Both of these are convoluted in a remarkable manner in the cavity of the belly, and terminate in the large intestines by a valve, which prevents the return of their contents. The large intestines, including the colon and rectum, or straight gut, constitute the lower termination of the abdominal viscera, and are destined to serve as a storehouse for all that portion of the food which is of no use to the system, and which is usually known under the names of dung and excrement. The masticated food then is received by the gullet into the stomach, and is further reduced to a finer state of division. The mode in which this division or solution of food is executed has not yet been satisfactorily ascertained. An acid certainly makes its appearance in the stomach when food is present, but whether this acid takes any part in the digestion or solution is still disputed. During the digestion of vegetable food in pigs, whose stomachs bear a close resemblance to those of man, I have always found a volatile acid present in minute quantities, which corresponded with the properties of acetic acid; but it is the only acid which distils over from the liquor of the stomach at a temperature of 212°. The filtered liquid of the stomach, under such circumstances, contains no hydrochloric acid, but an acid which is either lactic, or corresponds very closely with it.* To ascertain if free hydrochloric acid was present in the fluid contents of the stomach, after being distilled for some hours till no more acetic acid came over, the residue was filtered, and divided into three equal portions. 1. To the first portion a solution of nitrate of silver was added, until a precipitate ceased to fall; pure nitric acid was then added, and the temperature raised to the boiling point. The precipitate was filtered, washed, and weighed. 2. The second portion was evaporated to dryness, and ignited : the residue was dissolved in water, and precipitated by nitrate of silver, nitric acid being added, and the solution boiled. 3. The third portion was exactly neutralized with caustic potash, evaporated, and ignited : the residue was dissolved in water, and precipitated by nitrate of silver. The results of these experiments are indicated in the following table in grains :—

---

* Phil. Mag., April, May, 1845. Lancet and Medical Gazette of same year.

Experi-ments.	Weight of Chloride of silver.	Weight of Chlorine.	Weight of Hydro-chloric Acid.
1	7·81	1·95	2·00
2	7·17	1·79	1·84
3	7·97	1·99	2·04

The difference between the first and second experiments indicated the amount of chlorine in union with ammonia. In the third experiment the potash displaced the ammonia, and hence the amount of chlorine was the same in the first and third experiments. I therefore infer that no free hydrochloric acid was present. Hence it appears probable that this acid is produced at the expense of the sugar or starch of the food, and it appears doubtful if any considerable quantity of acid is secreted, as is generally imagined, from the coats of the stomach. Corvisart tells us, that in a case where there was an aperture in the stomach the contents of that organ during digestion were neutral ; and I have found the contents of the stomach of a sheep during digestion of grass, and several hours after the food had been introduced, without either an acid or alkaline reaction. A strong argument, however, against the hydrochloric acid theory of digestion is derived from the circumstance of the food containing, in many instances, but an insignificant quantity of chlorides, a considerable portion of which is again thrown out with the dung. Hay made from rye grass, for example, contains often merely a trace of chlorine, while in barley, and other kinds of grain, it is often entirely absent. Now as it is obvious that the hydrochloric acid, if any were present in the stomach, must be originally derived from the food, the absence of such a constituent in many kinds of food renders its disengagement in a free state in the

3*

stomach so much the less probable. I regret, therefore, to be obliged to infer that the commonly received
view of digestion is scarcely admissible. It is perhaps
safer to conclude, that there is a deficiency of knowledge on this important subject; and that not only do
we require to possess a few facts additional before we
can be said to understand the process, but we want an
entirely new basis on which to found a theory of digestion. It seems highly probable, from my own observations, that the starch of food is converted into sugar, and
that this again passes into simpler forms, as alcohol,
perhaps, acetic acid, or lactic acid, by a kind of substitution so well explained by the theory of Dumas, and
finally into gaseous forms, as carbonic acid and vapor
of water, or after some such fashion as suggested by
Liebig. The difficulty lies in explaining how the albumen and fibrin become dissolved, and are thus prepared to be taken up in a liquid state by the lacteals.
What has been described as fermenting or digesting
principles, under the names of pepsin, gasterase, &c.,
are obviously albumen, &c. modified by the action of
solvents, and have thrown no light hitherto on the nature of the solvent power. The most superficial ob-
server must have noticed that digestion is something
more than a mere chemical action. Does not the fam-
ished man *feel* refreshed after eating, and does not the
pulse beat quicker when food has been swallowed?
There is, therefore, a nervous action induced, the nature of which it is only wise to admit we do not as yet
understand. But so remarkable is the influence of
even simple food on the nerves, when abstinence has
been practised for some time, that it may be interesting

to quote the following case, in which intoxication was produced by the stimulus of oysters alone.

In the well-known mutiny of the Bounty, Capt. Bligh was set adrift in boats with twenty-five men about the end of April, in the neighborhood of the Friendly Islands, and was left to make his way to the coast of New Holland in such a precarious conveyance. At the end of May they reached that coast after undergoing the greatest privations, the daily allowance for each man having been one twenty-fifth of a pound of bread, a quarter of a pint of water, and occasionally a teaspoonful or two of rum. Parties went on shore, and returned highly rejoiced at having found plenty of oysters and fresh water. Soon, however, " the symptoms of having eaten too much began to frighten some of us; but on questioning others who had taken a more moderate allowance their minds were a little quieted. The others, however, became equally alarmed in their turn, dreading that such symptoms (which resembled intoxication) would come on, and that they were all poisoned, so that they regarded each other with the strongest marks of apprehension, uncertain what would be the issue of their imprudence !" Similar observations have been made under other circumstances. Dr. Beddoes states that persons who have been shut up in a coal-work from the falling in of the sides of a pit, and have had nothing to eat for four or five days, will be as much intoxicated by a basin of broth, as an ordinary person by three or four quarts of strong beer. In descending the Gharra, a tributary of the Indus, Mr. Atkinson states (Account of Expedition into Affghanistan in 1839–40, p. 66) that on two occasions during the passage he witnessed the intoxicating effects of food. To induce the Punjaubees

to exert themselves a little more, he promised them a
ram, which they consider a great delicacy, for a feast,
their general fare consisting of rice and vegetables made
palatable with spices.   The ram was killed, and they
dined most luxuriously, stuffing themselves as if they
were never to eat again.   After an hour or two, to his
great surprise and amusement, the expression of their
countenances, their jabbering and gesticulations, showed
clearly that the feast had produced the same effect as
any intoxicating spirit or drug.   The second treat was
attended with the same result.   The introduction of
food, therefore, into the stomach produces an influence
or sympathy over the whole body which is worthy of
notice, and shows us that we are too much disposed,
perhaps, to localize the physiological actions of the
systems of animals.

Digestive organs in animals which chew the cud.—
(*Ruminant animals, fig.* 2.)   The small and large intestines of these animals correspond, in general re-
spects, with those of the human subject.   The stomach
is, however, entirely different.   Instead of consisting
of one cavity as in men, the stomach of the sheep and
ox is divided into four compartments, which serve to
reduce the food to a finer state, and render it more
pulpy.

The food in these animals is first received into the
paunch, (*ventriculus,*) which occupies a large space in
the belly on the left side.   From this bag it passes into
the second stomach or honeycomb, (reticulum or bon-
net,) from the cell-looking aspect of its interior structure.   There the food is formed into a round ball, and
is thrown by the œsophagus into the mouth, to be again
chewed while the animal is at rest   This is termed

chewing the cud, and is a proof that the food has undergone little change in the first stomach. In the fine

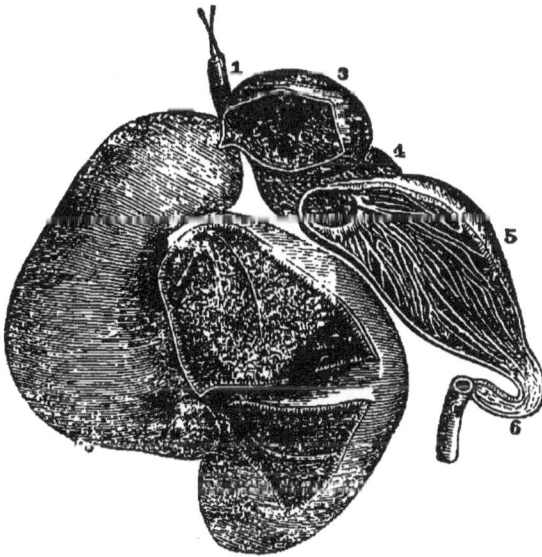

Fig. 2.

COMPOUND STOMACH OF RUMINANTS, (from Carus and Jones.)

1. Œsophagus.
2. The paunch, or first stomach.
3. The honeycomb, or second stomach.
4. The manyplies, or third stomach.
5. The caille or red, or fourth stomach.
6. The commencement of the small intestines.

state of division in which it now is, the food when
swallowed, "in consequence of its stimulating quality
being now altered, finds the two valvular folds at the
lower end of the œsophagus closed and shortened by
contraction, and is directed by the short canal they thus
form into the third, and thence into the fourth cavity of
the stomach," (*Grant*, p. 411,*) which is the true digest-

* Outlines of Comparative Anatomy, by R. E. Grant, M. D., &c.
Part IV. p. 410.

ing stomach, and is the one which is active when the young are suckling. The anatomy thus far at least of the ruminant animals is interesting to the cattle feeder, because it may explain the importance of mixing with grain a certain amount of chopped hay, in order that the whole may pass into the first stomach and have all the benefit of a second mastication; whereas, if it is administered at once in a fine state of division similar to that produced by chewing the cud, it may pass into the third stomach at once. The number of digesting operations to which vegetable food is thus subjected exhibits in a strong point of view the difficulty encountered by the systems of animals in extracting from this description of aliment the soluble ingredients fitted for their support. It is thus we find in man, that vegetable is longer of digesting than animal food, and that the American Indians, who live entirely on animals during a great portion of the year, are under the necessity of smoking largely the prepared bark of the willow to delay probably digestion, as the custom of smoking has been plausibly explained by Liebig. There is an interesting confirmation of the fact, if any were needed, of the easier digestibility of animal than of vegetable food, related in the case of Mr. Spalding, the improver of the diving-bell in the last century. He stated that when he had eaten animal food, or drunk fermented liquors, he consumed the air in the bell much faster than when he lived upon vegetable food and drank only water. Many repeated trials had so convinced him of this, that he constantly abstained from animal diet while engaged in diving. But as digestion is not confined to the stomach in the view which we have taken of it, we find that in animals

living on vegetable food the intestines are generally
much longer than in animals subsisting on animal food.
In the sheep, for example, they are twenty-eight times
the length of the body, while in animals which feed on
a mixed diet, the intestinal canal, as in man, possesses a
medium extent. • The importance of the length of this
tube is at once apparent for the digestion of a diet which
is with difficulty soluble, if we consider that the intestinal canal is believed to form an extensive surface, from
which the digested food is constantly passing away by
the mouths of vessels opening into it, termed lacteals.
These lacteals are considered to form a connection between the intestines and the bloodvessels, by which the
digested food, under the name of chyle, is transmitted into
the current of the blood. The chyle, which may therefore
be considered as incipient or young blood, contains similar ingredients to those which we find in the stomach,
viz., fibrin, albumen, sugar, oil, red coloring matter, and
salts. (*Prout.*) If we examine the blood when the chyle
has been mixed with it, we might expect to find indi-
cations of its presence in that fluid. Accordingly it
has been ascertained that the serum or watery part of
the blood, after partaking of a meal which contains any
fatty matter, is milky, and is not clear as is generally
supposed. This has been ascertained to be the case
in healthy men, and also in the inferior animals. For
example, calves were fed on gruel and milk, and after
various intervals they were slaughtered. The serum
of the blood on examination when the animal was killed
from three to six hours after the meal was found to be
milky, and to leave a greasy stain on filtering paper,
when the amount of milk or fatty matter used was
considerable; while the serum taken from an animal

which had been subjected to starvation for a space of time varying from twelve to twenty-four hours, presented generally a clear aspect.* Besides the fatty matter which had been used as food, traces of albuminous matter were detected in the serum of the blood when in the milky state; and from some experiments also it would appear that sugar, either derived from the starch, or from the saccharine matter of the food, can be detected in the blood. These observations, for an opportunity of making which I am indebted to Dr. A. Buchanan, seem to be corroborated by the fact stated by microscopical observers, that particles distinct from those of the fat can be detected in the chyle.

It has been a subject of discussion with physiologists, whether the chyle or incipient blood is taken up in the small intestines alone, or if absorption occurs also in the course of the large intestines. Upon this question it appears that no small degree of light may be thrown by a consideration of some circumstances in the feeding of cattle, which are sufficiently striking. As cows are continually feeding during the whole day, it can rarely happen that the stomach can be in any other condition than in that of engorgement, and yet the amount of water which the animals will swallow at a single draught is certainly more than sufficient to fill the whole of the cavities of the stomach supposing them to be empty. The following table will show the quantity of water swallowed by two cows on different occasions. The animals were placed on the weighing-machine, and their weight noted. They were then allowed to satisfy their thirst, and their weight was again taken.

---

* Paper by the author, Phil. Mag., April and May, 1845.

BROWN COW.

Food		Weight of Cow.		Water Swallowed.
		Before Drinking.	After Drinking.	
		lbs.	lbs.	lbs.
12 Aug.	Barley, molasses, and hay,	1010	1038	28
19 —	Malt and hay -	$998\frac{1}{8}$	1041	$42\frac{1}{8}$
29 —	Ditto - - -	$1023\frac{1}{2}$	$1048\frac{1}{2}$	25
4 Sept.	Barley, linseed, and hay,	991	1055	63

WHITE COW.

Food.		Weight of Cow.		Water Swallowed.
		Before Drinking.	After Drinking.	
		lbs.	lbs.	lbs.
12 Aug.	Barley, molasses, and hay,	1052	1106	54
26 —	Malt and hay -	1028	1051	23
4 Sept.	Barley, linseed, and hay,	1056	1104	48
13 —	Beans and hay -	1060	1087	27

In the fourth experiment with the brown cow, it will be observed that the animal swallowed at one draught sixty-three pounds weight of water. As the water was derived from the Clyde, and contained but a small amount of inorganic matter, we shall be very near the truth if we admit that the cow, on this occasion, swallowed six gallons of water without taking a breath. Now it is obvious that in these trials the water must have passed through the stomach into the intestines. On mentioning these facts to Sir Benjamin Brodie, to whose opinion in such experiments I most willingly defer, he informed me that he had found the water taken by small animals, when they were killed soon after swallowing it, to be lodged in the colon or large intestine. A similar observation has been made by Mr. Coleman, of the Veterinary College, in reference to the horse.—(*Bell.*) From which it has been inferred, that "the aliment is deposited liquid in the right colon; that in arriving in the rectum or straight gut, it is deprived of fluid, and that the lymphatics of the great intestine are found distended with a limpid fluid. From such views the idea has been entertained that a very principal office of the great intestines was to imbibe the fluid from their contents in proportion to the wants of the system."—(*Bell.*) It is not to be inferred, however, from the fact, that when the dung presents a less consistent aspect, it contains a much larger quantity of water. In the case of cows fed on grass, when the dung was thin and liquid, the percentage of solid matter was 11·27; while when they were feeding to a considerable extent on grain, and when the dung was very consistent, the amount of solid matter varied from 13 to 14½ per cent., affording evidence certainly of a greater quantity of water in the first instance than in the second, but not so considerable as might be expected from the external appearance of the substances.

If the view of Bell be correct, and it seems a very plausible opinion, the colon would appear to act the same office as the paunch and second stomach of the camel, dromedary, and llama, in which animals there are large cells in those portions of the stomach for the retention of water, which is thus supplied to the systems of the animals according to the exigencies of their case. Since the experiments which I have detailed appear to warrant the conclusion, that the water swa-lowed by the cows was conveyed into the colon. it is obvious that this water, in its passage through the stomach, must carry with it much soluble matter, especially of a saline nature, which may be absorbed through the coats of the great intestine, or thrown out with the excrementitious matter contained in the gut. It is in this way I am inclined to account for the considerable quantities of common salt and alkaline phosphates which I have met with in repeated analyses of the dung of cows fed on grass, hay, and grain. The amount of inorganic matter in cow-dung varies from 10 to 13 per 1000 parts; and in the latter case, the quantity of soluble salts, consisting of chlorides and phosphates, averaged as much as $1\frac{1}{4}$ per 1000 parts. The presence of these salts was quite unequivocal, as on burning the dung and digesting the residue in water the common salt was easily obtained in characteristic cubical crystals by concentration. The fact of the colon serving as a kind of reservoir for the large quantities of fluid carried into the intestinal canal, may serve also to explain the mode of action of saline purgatives. It would appear that, when dissolved in large quantities of water, they are carried at once to the colon, where they act by stimulating the intestine, increasing the peristaltic motion, and thus encouraging a more intimate mixture of the aqueous and solid contents of the gut, communicating the same liquid condition of the contents of this intestine to those of the rectum, which are usually quite free from water, and thus contributing to their easy evacuation. Liebig has endeavored to account for the action of saline purgatives by the power which they possess of extracting

water from the tissues, in the same way that common
salt extracts water from meat and forms brine. To a
certain extent this explanation is satisfactory; but it is
obvious it cannot extend to the action of powders, such
as jalap, &c., and accordingly Liebig restricts his view
to saline purgatives. But if, as Sir Charles Bell be-
lieves, there is always a quantity of water in the colon,
we can more readily understand how such vegetable
powders can act, and that their agency would be assisted by the use of diluents which will be carried
down to the rectum and be intermixed with its contents. The erect posture, if this view is correct, will
be the most proper to assume after the administration
of medicine, in order that the abundant draught of fluid
may be carried rapidly by gravity to the lower extrem-
ity of the intestinal canal. This explanation of the
action of purgatives, it will be observed, assimilates
them to clysters, with this difference, that a purgative
may act more or less from the stomach downwards,
while the influence of a clyster is generally restricted
to the rectum and colon. From this view we may also
infer, that, in cases where the bowels obstinately resist
the action of purgatives, and it is considered advisable
to administer a clyster, the action of the latter will be
facilitated by the free use of tepid water introduced by
the mouth. It may be further inferred from this view,
that a preference should be given to saline purgatives
over those of a vegetable nature, since, being soluble,
they are at once carried to the large intestines, their
proper sphere of action; and, contrary to the frequent
assertion, they are just as natural to the system as
those of a vegetable nature, since all wholesome food
contains saline ingredients. This view is, in some

measure, opposed to the employment of medicines in the state of pills, and would appear to dictate the propriety of administering aperients in the form of solution whenever it can be practised with propriety. This observation it is not intended, however, should be construed into a recommendation of the use of purgatives; on the contrary, we believe them to be much too frequently employed, and that a more intimate study of the process of digestion will convince both medical men and patients, that the primary object of attention is the nature of the food employed, and the due consideration of its adaptation to the particular circumstances in which an individual is placed. The nature of the action of purgatives now supported may be stated in a few words. The colon in a natural state contains water; the rectum contains only dry fæces : a purgative increases the action of the colon, intermixes the water and contents more intimately, propels these liquid matters into the rectum, occasions also a similar action to that induced in the colon, and finally, enables the whole contents to pass away with facility. This view is, in some measure, borne out by the fact of such succulent food as grass, which contains from $\frac{4}{5}$ to $\frac{2}{3}$ its weight of water, acting as an habitual aperient.

Purgatives are usually employed to remove, as the phrase goes, irritating matter from the intestines. Now, as the only foreign substance of any consequence, in addition to the food, thrown into the intestines, is the bile, it becomes an important object to determine upon what the physician is acting when he administers a purgative. The question, Where are the irritating materials lodged? demands first a solution. If in the colon, then why should the whole length of the intestinal canal be subjected to the stimulating action of a purgative, since the end can be more easily attained by throwing a clyster into the large gut? The second question is, Does the bile cause the irritation? And, third, Does not the food occasion the derangement? So little are we prepared to answer these questions, that we do not even as yet know the function or destination of the bile. But there can be little hesitation in affirming, that the use of purgatives is carried much too far in this country, especially mercurials, a class of the most dangerous poisons. The primary object of the introduction of food into the stomach and intestinal canal is to produce blood : in order that the latter may be of a healthy description, it is requisite that the food should contain the ingredients necessary for the production of blood, and that these should be in a state of integrity and health. It is scarcely to be wondered at that the consumption of putrid food, such as high-flavored game, and large quantities of decayed cheese, should be incapable of producing healthy blood ; or rather, that the blood produced from substances in such a state of putrefaction should be liable to disease of the most dangerous and deadly nature. One of the first considerations, then, in forming an opinion of the adequacy of food to produce healthy blood, is to compare its constituents with those of the blood. The true type of all food, as has been well demonstrated by Dr. Prout, is the milk which nature has provided so carefully for the use of sucking animals: in it we may expect to find all the substances requisite for the production of healthy blood. The following table affords, in parallel columns, a view of the ingredients entering into the composition of milk, wheat flour, and blood.

MILK.	FLOUR.	BLOOD.
	Fibrin.	Fibrin.
Curd or Casein.	Albumen.	Albumen.
	Casein.	Casein.
	Glutin.	Coloring Matter.
Butter.	Oil.	Fat.
Sugar.	Sugar, starch.	Sugar?
Chloride of potassium.		
Chloride of sodium.		
Phosphate of soda.		
Phosphate of lime.	Ditto.	Ditto.
Phosphate of magnesia.		
Phosphate of iron.		

From this table, therefore, we learn that the curd of milk is capable of undergoing certain modifications, which exhibit themselves under four forms in the blood. The coloring matter, too, of the blood is absent from the milk ; but the latter contains iron, which is connected with the coloring matter of the blood in some way not yet understood : and it was the opinion of Chaptal, and of others since his time, that the florid color of the blood was occasioned by the action of the oxygen of the atmospheric air upon the iron of the blood. But the experiments of Dr. Prout, who found a trace of coloring matter in the chyle, that is, in blood before it has been exposed to the action of the oxygen of the atmosphere, would appear to militate against this plausible view of the cause of the florid color of the blood ; and yet it is impossible to avoid the suspicion that further inquiry, and a more intimate acquaintance with the process of respiration, will connect, in some manner or other, the iron which exists in no other part of animals but the blood with the function of the oxidation of the systems of animals. But besides the necessity for the presence of the same materials in the food which exist in the blood, it is requisite that each should bear a certain relation to the whole, as will be attempted to be pointed out in the subsequent part of the work, during the discussion of the effects of the different kinds of diet employed in the extensive series of experiments to be detailed. The previous observations have shown the parallel nature of milk and blood. To make good milk, therefore, is obviously producing a similar effect to that of forming good blood, and consequently contributing to build up the body of animals in a healthy and substantial manner. Again, as the blood of cows is identical in composition with that of the human species, it is obvious that the diet of the one class of animals must possess a similar composition to that of the other. It is important, as a preliminary step, to consider briefly the nature of the animals upon which the experiments for determining the influence of different kinds of food as diet were made.

CHAPTER IV.

DESCRIPTION OF THE COWS.

DESCRIPTION OF BROWN AND WHITE COW.—INFLUENCE OF SYMMETRY UPON THE AMOUNT OF MILK.—THE HEALTH OF AN ANIMAL DEPENDS ON THE PROPER RELATION OF ITS ORGANS.—DIFFERENCE OF CONSTITUTION OF ANIMALS DEPENDS ON THE NERVOUS SYSTEM.—FAT ANIMALS OFTEN TO BE CONSIDERED AS IN A STATE OF DISEASE.

WHEN experiments are made upon a limited scale it is essential that the principal elements in the investigation should be carefully selected. Greater accuracy would be undoubtedly attained by experimenting upon a very large number of animals at the same time, provided that the execution could be effected with equal facility; but when the subsequent tables are examined, it will be at once evident that the labor, and consequent liability to error, attendant upon such researches when made in a more extensive form, would more than counterbalance any objections to a more limited scale of inquiry. In undertaking this series of experiments it was requisite to choose cows which should produce average results. The selection was intrusted to a very extensive agriculturist, (possessing a large herd of milk cows,) who was made acquainted with the object in view; and, from the results obtained, it appears that the choice was well made; and that, so far as the animals are concerned, there is probably nothing objectionable in the experiments. One of these animals was white or speckled, and the other was brown, and they answered to the following characters :—

White or speckled Cow.—This was a handsome cow of the Ayrshire breed, possessing a face of no great length, but of considerable breadth. The horns were curved inwards and forwards, and their tips turned slightly upwards. The neck was covered with patches of a brown color, and the rest of the body thinly spotted in the same manner. The spine formed a remarkably continuous horizontal line, unbroken by any depression. The chest was not characterized by a more than usual wedge-like form, although when viewed from behind, in connection with an expanded belly and short legs, this feature was to a certain extent observable. She therefore possessed undoubtedly an important element in a good milk cow, viz., large intestines and comparatively small lungs. This cow was five or six weeks calved, and had seen the bull a fortnight previous to the commencement of the experiments. The quantity of milk which she gave when at pasture, it was stated, was ten quarts, or about 25 lbs. 12 oz. imperial weight. This amount was never, however, reached during the whole course of the experiments, except upon one occasion. This animal was remarkably quiet; her age was between five and six years, and her weight, a fortnight after her arrival, 994 lbs.

Brown Cow.—This cow was considerably inferior in size to the preceding, and by no means endowed with a figure so pleasing to the eye of the connoisseur. Her horns protruded more. The spine was not straight, but was characterized by a decided dorsal depression, a mark of inferiority in an Ayrshire cow. Her color was brown, varied with a few white patches. Her belly did not protrude to such a degree as that of the white cow, and her lungs were in consequence larger in proportion. The quantity of milk which she gave at pasture is stated to have varied from nine to ten imperial quarts, a quantity which she much exceeded immediately after her arrival, but which gradually diminished and remained tolerably stationary till the close of the investigation. This cow had seen the bull two days before her arrival, but probably without the requisite effect, as she displayed occasionally considerable irritability, wildness of eye, and other well-known symptoms. The quantity of milk which she gave was generally less than that yielded by the white cow, but the amount of butter was greater. Her weight, a fortnight after her arrival, was 967½ lbs., and her age was about five years. She had calved five or six weeks.

It is not necessary, for the sake of elucidating the experiments, to discuss the much controverted points among agriculturists in reference to the form of cow best calculated for the purposes of the dairy, since practical judges differ as to the proper characters, and have too frequently fixed upon anatomical features as indicative of a good milk cow which are not necessarily so in a physiological point of view. No stronger proof could be adduced in support of this statement than the fact that the characters of a good milk cow of the short-horn breed are in many respects the reverse of those exhibited by the Ayrshire cow. The external symmetry of an animal must, in some measure, be viewed apart from its capacity to discharge a physiological function. It would be incorrect to judge of the capability of a man to undergo fatigue by the contour of his countenance, spine, and limbs alone, although their peculiar conformation might afford accessory proofs of power. Recent experiments, in accordance with scientific views, would tend to show that strength or endurance of fatigue will depend more upon the relation of one important division of the system to another, as of the organs of respiration, for example, to the stature or muscular development, than upon the general corporeal symmetry. A man of six feet and upwards may appear well proportioned to the eye, and yet experiment has shown that an inferior stature affords, on an average, greater muscular power, in consequence of the better ratio subsisting between the important organs which are necessary to the exercise of strength. This is at once obvious, if we bear in mind that the principal source of animal power is respiration, or that function by which certain portions of the digested food are converted into carbonic acid, acetic acid (?) and water; including, therefore, not only the lungs, but also the whole capillary system of the skin.* A short-winded person, or one whose respiratory organs are defective, is at once inferior in the capacity to undergo fatigue to another whose lungs are in a state of integrity; and this is the result, not merely because the lungs are somewhat diseased, but because, the exciting cause of all animal motion being dependent on the function of respiration,—that is, the conversion of carbon and hydrogen in the system into

---

* These views are strongly supported by the very ingenious experiments of Mr. Hutchinson, whose researches on respiration constitute a valuable contribution to physiology. Seo Journal of Statistical Society, June, 1844. Trans. of Med. Chirurg. Society of London, May, 1846.

carbonic acid and water,—it is requisite that the oxygen of the atmosphere should have access to a certain amount of blood-surface to produce a given effect. When any obstacle occurs to mar this operation,—for example, in consequence of disease of a portion of the lungs, or of the influence of a cause operating upon the whole constitution,—the inevitable result is a deterioration of muscular power. It is unnecessary to multiply examples in proof of the co-existence of muscular power and capacity of lung, since a broad chest is generally accepted as an element of strength. The relation between the muscles, or flesh, and the lungs being understood, it will be more easy to appreciate the connection between the intestines and the lungs. The intestines are the reservoir in which the food is placed for the purpose of being absorbed into the blood. The rapidity with which the dissolved or digested matter is taken up must, it is obvious, depend upon the rate at which the vessels destined for this purpose act; these being set in motion by the heart, this again by the nervous system, and the latter by respiration, there is discernible a beautiful chain of connection between the oxygen of the atmosphere and the absorbed food. If the system described were always in equable movement, if no influences were occasionally present to interfere with its proper equilibrium, animals would be in the condition of plants, which possess absorbing apparatus, but are destitute of one powerful interfering agent in the animal economy; this is the brain and nervous system, upon the condition of which depend passions and emotions of the mind. It is principally by the study of this important apparatus that we derive our knowledge of what is peculiarly termed the constitution of animals. Without this system animals would be merely chemical machines, and we might then predicate, in every case, the effects of particular influences, as one animal would then differ from another merely in the extent of its mechanism. The intestinal canal may then be considered as an extensive absorbing surface, which is retained *in equilibrio* by a properly-balanced exhaling surface, the lungs and skin. If there were no nerves, this equilibrium would spontaneously proceed, and every part of the animal system would be duly supplied with its proper amount of support. But to stimulate the nervous system we employ exciting substances, such as alcohol and spices, &c., which increase the rapidity of absorption without a corresponding provision being made for the proper exhalation of the excess of food thus introduced into the system. The consequence must be the deposition of fat, a condition of the system which is ranked in the human subject as a disease, (*Polysarcia adiposa.**) The same result occurs with the inferior animals if we force more food into their systems than can be in some degree proportionally exhaled. The deposition of fat ensues, and when it is carried to the extent too customary among agriculturists, it assumes the form of a disease : when cattle are fed for the purpose of serving as human food, there ought not to be such a superabundance of fatty matter deposited as is usual with some of the animal monsters designated fat cattle. When they are properly fed, with a due attention to allowing them a certain amount of exercise, the fat and

---

* In the language of Lord Byron, " fat is an oily dropsy."—*Reject ed Addresses*, p. 19.

lean are deposited in healthy proportions, and the cattle may be employed without risk as human food. Passions or mental influences must necessarily produce a decided effect upon the absorptive action of the intestinal canal, and may cause a diminished amount of nutriment to be absorbed: in this case the products of the animal, such as the milk of the cow, must necessarily be diminished. This remark is to be kept in view in considering the subsequent experiments. The cows were very different in reference to their nervous condition. The white cow was quiet and steady, generally eating equal portions and producing equable quantities of milk. The brown cow, on the contrary, was fitful in her appetite, and of consequence was variable in the amount of products. In proportion to her weight she consumed a larger amount of food than her fellow, but always afforded less milk and a greater amount of butter. The variable action of her organs is well exhibited in the first series of tables. When at pasture she had given two pints less than the white cow, and immediately before the experiments she gave the same quantity as her fellow. On her arrival in Glasgow her milk greatly increased; but it soon began to diminish, although the same amount of food was continued. That the change was not produced by any alteration in the food is obvious from the steadier result afforded by the white cow, which was also supplied with an equal weight of fodder. The amount of milk given by the brown cow was as much as 26 lbs per day when she was fed with grass, and upon the same kind of food the quantity declined to 22 lbs.; while the milk produced by the white cow was, at the commencement of the experiment with grass, 23

lbs., and at the termination of the trial, 21 lbs.; so
that there was a falling off, in the case of the brown
cow, to the extent of 4 lbs., and with the white cow
only to the amount of 2 lbs.   That this result was not
merely owing to a deficiency of water was proved by
experiment, which gave the same amount of water in
the milk of both cows; but the quantity of butter afforded by the brown cow amounted to $11\frac{1}{4}$ lbs., while
that of the white cow was $8\frac{1}{8}$ lbs., in fourteen days,
from 1,427 lbs. of grass supplied to each animal
Again, when the animals were fed on steeped entire
barley, the brown cow's milk fell from $22\frac{1}{2}$ lbs. to $17\frac{1}{2}$
lbs., while that of the white cow only declined from
22 lbs. to $19\frac{1}{2}$ lbs.; the brown cow falling off to the
extent of 5 lbs., and the white only to the extent of
$2\frac{1}{2}$ lbs.   These facts are sufficient to show that the
two animals were constitutionally different.   The oc-
casional wild look of the brown cow, her tendency to
gore those who approached her, her frequent startled
aspect, all indicated a nervous state of excitement; the
probable cause of which has been already alluded to.
The result of these experiments seems to countenance
the idea, that, although a handsome external figure is
not necessarily an indication of the highest capacity in
a cow to produce milk and butter, yet that it may conduce to afford a steady supply of milk, inasmuch as it
appears to indicate a proper relation between the organs.

   Color of Cattle.—It has been supposed by some
practical persons that the color of an animal exercised
some influence on the amount of milk produced.   The
determination of this point could only be decided by
experiments upon different breeds of cattle; but it is

probable that color is not an important element in this inquiry, any further than that the same parents being good milkers may originate a stock of similar character, both in color and in functions, to themselves; and hence a particular color co-existing with good milking capacity would rather be an accidental than a physiological circumstance. The subject is one, however, open for inquiry, and is alluded to here because it is a favorite idea with some good practical observers.

In the experiments to be detailed, it is proper to state that the milk was carefully weighed and also measured morning and evening; the numbers contained in the series of tables are therefore the exact results of experiments. The weight of grain may be taken as representing the exact chemical quantities, while the amount of hay being only given in quarter pounds might be received as the practical quantities, and not as the precise chemical numbers. The dung was also carefully weighed morning and evening, and its solid and liquid contents estimated by frequent desiccations. The butter was extracted from the whole of the milk. The morning's milk was allowed to stand for twenty-four to thirty-six hours, and was then creamed; the cream being placed in the churn, together with the whole of the evening's milk. The weights and measures used are all Imperial.

5*

CHAPTER V.

INFLUENCE OF GRASS WHEN USED AS DIET.

TABLES OF MILK AND BUTTER PRODUCED BY GRASS DURING FOURTEEN
DAYS.—COMPOSITION OF THE MILK.—AMOUNT OF FOOD CONSUMED.—
OF THE SOURCE OF THE BUTTER IN THE GRASS.—AMOUNT OF WAX
IN THE FOOD.—COMPOSITION OF BUTTER.—MODE OF PRESERVING BUT-
TER FRESH FOR ANY LENGTH OF TIME.—IMPROBABILITY OF WAX BEING
CONVERTED INTO BUTTER.—ON THE NATURE OF GRASS AND HAY AS
FOOD.—ANALYSIS OF HAY.—GRASS LOSES NUTRITIVE MATTER WHEN
CONVERTED INTO HAY IN THIS COUNTRY.—TABLE OF FALL OF RAIN.
—PROCESS OF ARTIFICIAL HAYMAKING SUGGESTED.—ANALYSIS OF STEM
AND SEEDS OF RYE-GRASS.—IMPORTANCE OF MAKING HAY BEFORE GRASS
BEGINS TO SEED.

IMMEDIATELY before the commencement of this experiment, the cattle had been grazing, and were brought
a distance of about forty miles by railway; a circumstance which may account for several irregularities and
anomalies in the immediate subsequent history of the
animals as derivable from the tables :—

EXPERIMENT I.—GRASS DIET.

BROWN COW.

Days.	Date.	Milk. lbs. oz. drs.	Grass. lbs. oz. drs.	Dung. lbs. oz. drs.	Weight of Cow. lbs.	Butter. lbs. oz. drs.	Temp.
	1845:						
1	June 10	26 11 4	93 5 13	79 13 11	...		57°
2	— 11	27 3 3	93 5 15	77 15 7	...		62
3	— 12	24 14 12	100 0 6	67 1 12	...	3 15 0	69
4	— 13	26 12 15	100 0 0	68 8 9	...		68
5	— 14	26 4 9	100 0 0	66 14 0	...		67
6	— 15	25 5 11	100 0 0	76 0 11	...		63
7	— 16	26 3 8	100 0 0	78 3 11	...		59
8	— 17	24 12 11	100 0 0	71 14 8	...	4 6 4	65
9	— 18	24 0 1	100 0 0	59 12 7	967½		62
10	— 19	22 2 1	120 0 0	64 8 10	...		59
11	— 20	21 5 7	100 0 0	74 6 12	...		62
12	— 21	22 9 4	120 0 0	78 11 15	...	2 15 0	60
13	— 22	21 12 10	100 0 0	87 7 14	...		57
14	— 23	22 12 1	100 0 0	97 10 3	986		58
		342 14 1	1426 12 0	1049 2 2	Gain, 18½	11 4 4	

EXPERIMENT I.—GRASS DIET.

WHITE COW.

Days	Date	Milk (lbs. oz. drs.)			Grass (lbs. oz. drs.)			Dung (lbs. oz. drs.)			Weight of Cow (lbs.)	Butter (lbs. oz. drs.)			Temp.
1	1845: June 10	23	4	14	93	5	13	70	9	8	…				
2	— 11	21	10	5	93	5	15	70	0	5	…				
3	— 12	22	1	2	100	0	0	70	3	0	…	3	2	4	3,45
4	— 13	23	6	8	100	0	0	62	4	9	…				
5	— 14	25	15	12	100	0	-0	67	11	13	…				
6	— 15	21	14	13	100	0	0	66	8	1	…				
7	— 16	22	5	5	100	0	0	72	2	5	…	2	15	12	2,25
8	— 17	20	3	3	100	0	0	62	4	14	…				
9	— 18	22	3	5	100	0	0	74	3	15	994				
10	— 19	20	0	15	120	0	0	66	1	3	…				
11	— 20	18	12	1	100	0	0	61	3	9	…	2	0	0	2,35
12	— 21	20	14	0	120	0	0	79	10	15	…				
13	— 22	20	11	5	100	0	0	86	9	8	…				
14	— 23	21	5	10	100	0	0	90	13	0	1044				2,67
		304	13	2	1426	12	0	1000	7	9	Gain, 50	8	2	0	

Proximate Analysis of the Experiment.—The com position of the grass, consisting almost entirely of rye grass, (*Lolium perenne,*) and of the dung, was as follows :—

	Grass.	Dung.
Water	75·	88·33
Sol. Salts	} 1·34	{ 0·40
Silica and Insol. Salts		1·35
Organic Matter	23·66	9·92
	100·	100·

Hence the solid matter in the food of the brown cow was 356 lbs., in the dung 147, while in the food of the white cow there were 356 lbs. of solid matter, and in the dung 140 lbs., making in all 425 lbs. swallowed by the two cows.

The composition of the milk of the cows was as follows :—

	Brown.	White.
Spec. Grav.	1029·8	1029·8
Water	87·19	87·35
Butter	3·70	
Sugar	4·35	
Casein	4·16	
Sol. Salts	0·15	0·156
Insol. Salts	0·44	0·488

From the previous experiments it therefore appears, that the same quantity of food given to cows nearly of the same weight produced 5 lbs. less of solid matter of milk in one cow than in the other ; 100 lbs. of solid matter of grass producing in the brown cow 17½ lbs

of dry milk, and in the white cow only 15½ lbs.   From
the column, however, in which the weight of the cattle
is represented, it appears that both cows were increasing in weight ; but, as the white cow advanced most
rapidly, it is probable that the difference in the quantity
of solid milk may have been applied to increase the
weight of the white cow.   There is another alternative
which is also admissible, viz., that the capacity of the
lungs and respiratory organs of the white cow were
greater than those of the brown cow, since the former
absorbed a greater amount of solid matter from the
grass, as appears from the difference between the grass
and dung, than in the case of the brown cow.   These
important differences in the two animals rendered it
impracticable to make comparative experiments upon
them at the same time.   The only method which could
afford results of value was, to supply each with the
same kind of food, and thus to obtain data which could
enable a judgment to be formed of the relative nature
of the constitutions of the animals.

The whole series, therefore, consists of two parallel
sets of experiments, the second of which may be viewed
as a repetition of the first trials, thus serving to control
any liability to error which might readily occur from
the nature of the investigation.

Ultimate Analysis of the Experiment.—The ultimate composition of the grass and dung was found to
be as follows :—

	Grass.		Dung.	
	Fresh.	Dried at 212°.	Fresh.	Dried at 212°.
Carbon - -	11·35	45·41	6·40	45·74
Hydrogen -	1·48	5·93	0·78	5·64
Nitrogen - -	0·46	1·84	0·25	1·81
Oxygen -	10·39	41·54	5·20	37·03
Ash - - -	1·32	5·28	1·37	9·78
Water -	75·00		86·00	
	100	100	100	100

Table exhibiting the Amount in Pounds of Carbon, &c. in the Food and Dung during Fourteen Days.

	BROWN.			WHITE.		
	Grass.	Dung.	Consumption.	Grass.	Dung.	Consumption.
Carbon -	161¾	67	94¾	161¾	64	97¾
Hydrogen -	21	8	13	21	7¾	13¼
Nitrogen -	6½	2$\frac{7}{10}$	3$\frac{9}{10}$	6½	2½	4
Oxygen -	148	54½	93½	148	52	96
Ash - -	18¾	14⅓	4$\frac{4}{10}$	18¾	13¾	5
Water - -	1070¾	902½	167½	1070¾	860	210¾
	1426¾	1049	377	1426¾	1000	426¾

From this table we learn that the brown cow consumed daily 6¾ lbs. of carbon; this is very nearly equivalent to 1 oz. of carbon for every 9⅓ lbs. of live weight, (the cow weighing 8 cwt. 71 lbs.) The white cow consumed daily nearly 7 lbs. of carbon, or 1 oz. of carbon to 8¾ lbs. of live weight; and the daily consumption of all constituents is represented in the following table, which affords a view of the mean of the two cows :—

				lbs.
Carbon	-	-	-	6·87
Hydrogen	-	-	-	0·93
Nitrogen	-	-	-	0·28
Oxygen	-			6·76
Ash	-	-	-	0·33
Water				13·50
				28·67

That so much matter should be ejected by animals is a circumstance liable to excite surprise in one who examines the physiology of digestion merely in a cursory manner ; but when we recollect that the stomachs of a ccw are of great capacity, capable of holding several gallons of water, and that these vessels, if we may so speak, require to be filled, in order that a mechanical excitement may be communicated to their surrounding coats, we may discover perhaps why a condensed regimen, although it might contain sufficient nourishment to supply the waste of the body, from its insufficiency of bulk to excite the stomach to secrete the requisite gastric fluid, might be incompletely digested. Hence it may be that grain and all farinaceous food are insufficient for cattle : they require a quantity of hay or straw in addition, for the purpose, in common language, of filling up the animal, but possibly to excite the coats of the stomach to the action of secretion. It is perhaps a prefe able view to consider the hay as containing a larger amount of calorifient constituents.

Of the Constituent of the Grass which supplies the Butter.—It is now upwards of a century since Beccaria of Bologna broached the idea that animals are composed of the same substances which they employ as food :—
" En effet si l'on excepte la partie spirituelle et immor-

telle de notre être, et si nous ne considérons que le
corps, sommes nous composés d'autres substances que
de celles qui nous servent de nourriture. (1742.)"—
Collection Académique, tome x. p. 1.   In more recent
times Dr. Prout has defended the same doctrine, and
has referred us to milk as the type of nourishment.
In this fluid the main solid constituents are oil, fibrin,
and sugar; these, therefore, or analogous bodies, he
considers should enter into the composition of all wholesome nutriment.   Still more lately a difference of opinion has resulted with reference to the exact part which
starch or sugar plays in the animal economy.   Fibrinous
matters, it is generally admitted, undergo little or no
alteration in the system; but whether it is necessary,
in order to produce fat in an animal, that the food should
contain oil, and that no other form of nutriment can
produce this substance, is a question which has been
very much debated.   It has been contended that the
presence of oil, if not essential in the food, is at least
very important in increasing the amount of fat deposited; while Liebig holds, that oil may possibly be assi-
milated or converted into butter, but that the same product may result from the deoxidation of starch or sugar
in the animal economy.   To the agriculturist the settle-
ment of this question is of no small importance, since it
may guide him to the use of various kinds of food for
the fattening of cattle which may otherwise be overlooked, and may also conduce to the proper prepara
tion of food, a subject which has received less attention
than perhaps it deserves.   In the prosecution of the
present series of experiments the prospect of throwing
some light upon this interesting subject has been kept
in view; and, in general, such experiments as were

required to afford data for calculating, from the different
kinds of food, the probable origin of the oily matter
secreted by the animals, have been carefully registered.
To solve the question, it is necessary to ascertain the
amount of oil in the food. The oily matter in the
grass was determined by first drying the grass at the
temperature of 212°, to remove water; it was then
digested in successive portions of ether, until this liquid
ceased to remove any matter in solution. The same
experiment was performed with the dung. The first
process, therefore, gave all the oily matter swallowed
by the animal, and the second afforded the oil or wax
which was not taken into the system: 2000 grains of
grass, when dried, became 500 grains. By digestion
in ether, 42·3 grains were taken up of a matter having
a dry waxy consistence, possessing a green color, but
without any of the characters of a fluid oil; this is
equal to 2·01 per cent. 4284 grains of moist dung
from grass, equivalent to 500 grains of dry dung, afforded 13·2 grains of an exactly similar green waxy
matter to that found in the grass, equal to 0·312 per
cent. The largest amount of wax in the dung of the
cattle was obtained while they were feeding on hay;
1000 grains of dung left, at the temperature of 212°,
157 grains of dry dung, which gave 6 grains of wax,
equivalent to 0·6 per cent. in moist dung, or 3·82 per
cent. in the dry dung. All of these products were
carefully dried for some days at the temperature of
boiling water. From these data, then, we are enabled
to construct the following table :—

	lbs.
Amount of wax in food of both cows in fourteen days	57·3
Amount of wax in dung -	6·3
Amount of wax consumed by the cows -	51·0
Amount of dry butter - -	16·7
Excess of wax in the food - - -	34·3

To ascertain whether the whole of the butter is removed from the milk by the usual process of churning, portions of the same milk were analyzed by the usual methods, for the sake of comparison. The brown cow's milk in the present experiment contained 3·46 per cent. of butter, while, by analysis, the amount was 3·7, making a difference of rather less than a quarter of a pound in 100 pounds of milk. This is so small that it does not affect the preceding calculation, but rather tends to show that the determination of such questions on a large scale is preferable to the usual analytic methods, since the analysis of milk twice a day for several months would be such a laborious work as to render its accomplishment impossible.

It is necessary to explain the circumstance that butter, as obtained by the usual mechanical process, contains foreign matter, consisting of water and curd, or casein. By analysis, butter was found to have the following composition :—

Casein	-	-	0·94
Oil	-		86·27
Water	-		12·79.

The composition of French butter has been stated to be somewhat different, (Boussingault,) as it has been found to contain upwards of eighteen per cent. of impurity. This difference may be owing to the coldness of the summer during which the present experiments were made.

The hardness of the butter was a subject of general remark, and might render it better fitted for being freed from the casein than if it had possessed a more fluid form.

Mode of preserving Butter fresh.—The cause of the tainting of fresh butter depends upon the presence of the small quantity of curd and water as exhibited by the preceding analysis. To render butter capable of being kept for any length of time in a fresh condition, that is, as a pure solid oil, all that is necessary is to boil it in a pan till the water is removed, which is marked by the cessation of violent ebullition. By allowing the liquid oil to stand for a little the curd subsides, and the oil may then be poured off, or it may be strained through calico or muslin, into a bottle, and corked up. When it is to be used it may be gently heated and poured out of the bottle, or cut out by means of a knife or cheese-gouge. This is the usual method of preserving butter in India, (ghee,) and also on the Continent; and it is rather remarkable that it is not in general use in this country. Bottled butter will thus keep for any length of time, and is the best form of this substance to use for sauces.

From the preceding table it appears, that the oil consumed by the cows greatly exceeded the butter, and the oil contained in the dung, even if the casein and the water were not subtracted from the butter; the total quantity of butter being 19 lbs. 6 oz. The result of this experiment is in perfect accordance with the facts observed by Boussingault, who, in similar researches upon cattle, found the oil in the food to exceed that in the dung and milk. The matter extracted by ether from grass, however, can scarcely be termed an oil, since it possesses all the characters of a wax; that is, a body which contains a smaller amount of oxygen than a fat oil,—certainly less than is contained in butter. It is therefore difficult to conceive a wax to obtain more oxygen in the system, and to be converted into an oil, where all the actions are calculated to remove oxygen, and not to supply it : such an occurrence would be as probable as the addition of oxygen to wood by throwing it into a furnace. The production of butter from sugar by the action of casein or curd is, on the contrary, a process with which chemists are now familiar, and is therefore more readily admissible into physiological theories than the idea of the formation of butter from wax, since we are unacquainted with any analogous example. The connection between sugar, oil, and wax is exhibited by the following formula :—

	Carb.	Hyd.	Oxyg.	*Differences.*		
				Carb.	Hyd.	Oxyg.
Sugar -	48	44	44			
Fat -	44	40	4	4	4	40
Wax	40	40	2	4	0	2

In bees we have a well demonstrated example of the production of wax from sugar, while fat, or the intermediate stage, is probably first produced in the body of the bee, and is then, by the loss of a small portion of carbon and oxygen, converted into wax, or to the lowest state of oxidation existing in the animal system. The point therefore to which it is necessary to direct attention is, that we have instances in chemical physiology of substances being produced from the others preceding it in the table, but that we are unacquainted with any phenomena of an inverse order ; nor would such an occurrence be explicable upon the principles on which the animal system is understood to proceed. Taking all these circumstances into consideration, it appears that there are fewer difficulties in the way of supposing that butter is formed from the starch and sugar, or albuminous matter, of the food, than from the waxy matter which is present in such considerable quantities. There is only one instance, with which physiologists are at present acquainted, that could be adduced as evidence in favor of any substance being rendered more complex in the animal system, viz., the production of fibrin or flesh from curd or casein. So far as chemical experiments carry us, we are not in a condition to affirm that no fibrin exists in milk, but it is admitted that none has as yet been detected. If these be correct, then it would appear to follow that the infant fed on milk must derive its flesh from the curd of that fluid, and that as curd contains no phosphorus, (while fibrin does,) the curd of the milk, in order to form muscular fibre, is united to phosphorus in the animal system, and is thus built up, instead of being, as is the rule with other substances, reduced to a smaller number of elements.

The objection to this view of the subject is, that the experiments which have been made on fibrin do not prove that it contains phosphorus ; they only prove that phosphoric acid can be detected in it even when it is purified in the most careful manner suggested by chemical knowledge ; and it would therefore be somewhat premature to adopt any such analogy as that which we have been considering.*

---

* When this passage was written, in November, 1845, I founded

On the Nature of Grass and Hay as Food.—Grass, as may be readily imagined, varies very considerably in its composition, according to its age, and also, as may be expected, according to its species. The experiments undertaken during the present investigation have sufficiently demonstrated the first of these positions; but the second is still open for inquiry, since chemists who have previously analyzed grass and hay have omitted to particularize the botanical names of the plants which they have examined. The grass used in the present experiments consisted almost entirely of rye grass, (*Lolium perenne*,) and the hay employed was also similarly constituted.

It may be interesting, for the sake of comparison, to give a table of the analysis of such specimens of hay as have been analyzed hitherto :—

my reasouing in reference to the probability of phosphorus not being a constituent of animal substances partly on the circumstance that Fremy, in his analysis of the acid of the nerves, (cerebric acid,) found 0·9 per cent. of phosphorus ; while, in my examination of the same substance, further purified, I found only 0·46 per cent. Since that period, however, Liebig has found that, when properly prepared, fibrin and albumen are destitute of phosphorus. In the May number of the Philosophical Magazine for 1846, I have described a modification of fibrin under the name of *pegmin*, well known as the buffy coat of inflamed blood. This substance contains sulphur, and cannot therefore be termed an oxide of protein. Under the name of pyropin I have also described a ruby-colored substance found in the position of the pulp of the elephant's tooth. The following is their composition :—

	Pegmin.	Pyropin.	
		I.	II.
Carbon	52·07	53·33	53·50
Hydrogen	7·00	7·52	7·66
Nitrogen	14·31	14·50	38·84
Oxygen	26·62	24.65	
Sulphur			

I. Analysis of hay made at Giessen by Dr. Will the species of grass is not mentioned.

II. Hay grown in the neighborhood of Strasburg in France, analyzed by M. Boussingault: the name of the grass is omitted.

III. Analysis of *Lolium perenne*, as previously given and used in the present experiments.

	I.	II.	III.
Carbon -	45·87	45·80	45·41
Hydrogen	5·76	5·00	5·93
Nitrogen		1·50	1·84
Oxygen	41·55	38·70	39·21
Ash - -	6·82	9·00	7·61

Although the species of grasses constituting these specimens of hay were in all probability different, the correspondence in their composition is sufficiently striking.

The amount of solid matter in this grass varied from eighteen to upwards of thirty per cent., according to the early or late period of its growth. The grass made use of in the first experiment contained from eighteen to twenty-five per cent. In our calculations the latter number has been adopted.

When grass first springs above the surface of the earth the principal constituent of its early blades is water, the amount of solid matter being comparatively trifling; as it rises higher into day the deposition of a more indurated form of carbon gradually becomes more considerable; the sugar and soluble matter at first increasing, then gradually diminishing, to give way to the deposition of woody substance.

The following table affords a view of the composition of rye-grass before and after ripening :—

	18th June.	23d June.	13th July.
Water -	76·19	81·23	69·00
Solid Matter -	23·81	18·77	31·00

These are important practical facts for the agriculturist; for if, as we have endeavored to show, the sugar be an important element of the food of animals, then it should be an object with the farmer to cut grass for the purpose of haymaking at that period when the largest amount of matter soluble in water is contained in it. This is assuredly at an earlier period of its growth than when it has shot into seed, for it is then that woody matter predominates ; a substance totally insoluble in water, and therefore less calculated to serve as food to animals than substances capable of assuming a soluble condition. This is the first point for consideration in the production of hay, since it ought to be the object of the farmer to preserve the hay for winter use in the condition most resembling the grass in its highest state of perfection. The second consideration in haymaking is to dry the grass under such circumstances as to retain the soluble portion in perfect integrity. To ascertain whether hay, by the process and exposure which it undergoes, loses any of its soluble constituents, the following experiments were made :—

1st.—3000 grains of rye-grass in seed on the 13th July, gave up to hot water a thick sirupy fluid, which, when dried till it ceased to lose weight at 212°, weighed 217·94 grains, equivalent to 7·26 per cent.

2d.—2500 grains of rye-grass, digested in cold water, yielded 53·23 grains of extract, equal to 2·12 per cent. This rye-grass contained 31 per cent. of solid matter, and 69 per cent. of water.

3d.—New hay, made from rye-grass, and containing 20 per cent. of water, for the sake of comparison, was also subjected to similar trials.

	Grains.	Grains.	
1st. 1369 gave to hot water	220·77 of extract,	16·12 per cent.	
1000	159·34	15·93	
1000	140	14	

2d. 2000 grains of new hay, in seed, digested in cold water, yielded 101·3 grains of extract = 5·06 per cent. of soluble matter.

From these numbers we learn that 100 parts of hay are equivalent to 387½ of grass. This amount of grass should contain of soluble matter in hot water 28·13 parts, and in cold water 8·21 parts. But the equivalent quantity of ·hay, or 100 parts, only contains 16 instead of 28 parts soluble in hot water, and 5·06 instead of 8¼ parts soluble in cold water. A very large proportion of the soluble matter of the grass has obviously disappeared in the conversion of grass into hay. The result of the haymaking in this particular instance has, therefore, been to approximate the soft, juicy, and tender grass to woody matter, by washing out or decomposing its sugar and other soluble constituents. These facts enable us to explain the reason why cattle consume a larger amount of hay than is equivalent to the relative quantity of grass. Thus ani-

mals which can subsist upon 100 lbs. of grass should
be able to retain the same condition by the use of 25
lbs. of hay, if the latter suffered no deterioration in
drying. The present series of experiments, however,
show that a cow, thriving on 100 to 120 lbs. of grass,
required 25 lbs. of hay, and 9 lbs. of barley or malt,
affording thus collateral evidence of the view which we
have taken of the imperfection of the process of haymaking at present in use in this country.

The great cause of the deterioration of hay is the
water which may be present, either from the incom-
plete removal of the natural amount of water in the
grass by drying, or by the absorption of this fluid from
the atmosphere. Water when existing in hay from
either of these sources will induce fermentation, a process by which one of the most important constituents
of the grass,—viz., sugar—will be destroyed. The
action necessary for decomposing the sugar is induced
by the presence of the albuminous matter of the grass ;
the elements of the sugar are made to re-act on each
other in the moist state in which they exist, in consequence of the presence of the water and oil, and are
converted into alcohol and carbonic acid according to
the following formula :—

	Carb.	Hyd.	Oxyg.
1 atom sugar - - - -	12	12	12
2 atoms alcohol - - -	8	12	4
4 atoms carbonic acid - - -	4	0	8

That alcohol is produced in a heated haystack in
many cases may be detected by the similarity of the
odor disengaged to that perceptible in a brewery. We
use this comparison because it has been more than

once suggested to us by agriculturists. The quantity of water or volatile matter capable of being removed from hay at the temperature of boiling water varies considerably. The amount of variation during the present experiments was from 20 to 14 per cent. If the lower per-centage could be attained at once by simple drying in the sun, the process of haymaking would probably admit of little improvement; but the best new-made hay that we have examined contained more than this amount of water, the numbers obtained verging towards 20 per cent. When it contains as much as this it is very liable to ferment, especially if it should happen to be. moistened by any accidental approach of water. The only method which we have found to succeed in preserving grass perfectly entire is by drying it by means of artificial heat. Rye grass contains, at an early period of its growth, as much as 81 per cent. of water, the whole of which may be removed by subjecting the grass to a temperature considerably under that of boiling water; but, even with a heat of 120°, the greater portion of the water is removed, and the grass still retains its green color, a character which appears to add greatly to the relish with which cattle consume this kind of provender. When this dried grass (as it may be truly termed by way of distinction from hay) is examined, it will be found to consist of a series of tubes, which, if placed in water, will be filled with the fluid, and assume in some measure the aspect of its original condition. In this form cattle will eat it with relish, and prefer it to hay, which, in comparison, is blanched, dry, and sapless. The advantages obtained by this method of making hay, or rather of preserving grass in a dry state, are sufficiently obvious. By this means all the constituents of the grass are retained in a state of integrity; the sugar, by the absence of water, is protected from undergoing decomposition, the coloring matter of the grass is comparatively little affected, while the soluble salts are not exposed to the risk of being washed out by the rains, as in the common process of haymaking. The amount of soluble matter capable of being taken up by cold water is, according to the preceding trials, as much as 5 per cent., or a third of the whole soluble matter in hay. We may therefore form some notion of the injury liable to' be produced by every shower of rain which drenches the fields during hay harvest. It is not only, however, the loss which it sustains, in regard to the sugar and soluble salts, that renders hay so much less acceptable than grass to the appetite of cattle. The bleaching which it undergoes in the sun deprives it of the only peculiarity which distinguishes the one form of fodder from the other; grass deprived of its green coloring matter presents exactly the appearance of straw, so that hay ought to be, termed grass straw. It is obvious, from the experiments detailed, that the operation of haymaking, as conducted in this country, has a tendency to remove a great proportion of the wax in the grass. Thus it was found that rye-grass contained 2·01 per cent. of wax. Now as $387\frac{1}{2}$ parts of rye-grass are equivalent to 100 parts of hay, and as $387\frac{1}{2}$ parts of grass contain 7·78 parts of wax, it is obvious that 100 parts of hay should contain the same amount of wax; but by experiment it was found that 200 grains of hay contained 4 grains of wax, which is equivalent to 2 per cent., almost exactly the amount contained in grass. Hence it appears

7

that no less than 5·78 grains of wax have disappeared during the haymaking process. The whitening process which the grass undergoes in drying renders it apparent that the green coloring matter has undergone change; but that it should have been actually removed to such an extent, or at least have become insoluble in ether, is a result which could scarcely have been anticipated without actual experiment. Some improvement in the preparation of hay is imperatively demanded in such localities as are affected with a more than usual fall of rain. The following table of the fall of rain will point out where such precautions are more particularly required :—

	Inches.	
Glasgow - -	21·3	
London - -	24·0	
Edinburgh	24·5	
Berwickshire - - -	32·5	{ Abbey St. Bathans, 400 feet above sea.
Manchester	36·1	
Lancaster -	39·7	
Paisley - -	47·1	at the Reservoir.
Strathaven -	45·8	700 feet above sea.
Greenock	61·8	{ 800 feet above the town.

The Glasgow result is the mean of many years' observation at the Macfarlane Observatory. The London is taken from the Royal Society Register, the mean of ten years. The Edinburgh number is from observations at the observatory. The Berwickshire number is the mean of two years' register, by Mr. Wallace, kept at my request. The Manchester and Lancaster are from Dr. Dalton. The Paisley and Greenock results are from the water-works register, the mean of seven years. The Strathaven number is from registers kept at my request by Mr. Wiseman.

Frequently the quantity of rain which falls in May and June, the haymaking season, is greater than in April and July. In those localities where the fall of rain is so considerable, the preparation of good sound hay by the usual process will be almost impracticable, and in such places too frequently hay in a state of decomposition is given to animals, at the risk of their being seriously injured, since all food whose particles are in a state of fermentation or putrefaction, which are analogous actions, must have a tendency to produce similar decompositions in the fluids of the animal system. In the neighborhood of manufacturing towns there could be no difficulty in preparing abundance of hay by the process now recommended. The waste heat of the chimneys might be sent through apartments or sheds of almost temporary construction, guided by a proper draught, so as to carry off the vapor as soon as it is volatilized; and the same arrangements might, with economy, be adopted in conjunction with brick and tile works. Haymaking would thus commence at a much earlier period of the season, the grass would be cut, carted to the drying-room, and in the course of a few hours be ready for stacking. When hay prepared in this manner is to be given to cattle and horses it may be steeped in a tank for twenty-four hours, or any adequate period, before being placed in the racks and boxes; and the steep water, which will contain sugar and soluble salts, should be given them to drink.

By this system of preserving grass we should be continuing to our cattle in winter our summer food, which all admit to be superior to every other substitute; and while the animals themselves would be benefited, much uneasiness and trouble in winter would be saved to the farmer. In a moist climate, especially like that exhibited in Scotland during the last year, it appears highly desirable that farmers should possess on their premises a drying-room, where hay, potatoes, and even corn, might be dried. Had such a convenience been attached to many of our farmers' offices last season much corn might have been saved, even by drying one or two cart-loads daily. This desideratum might be effected by running a flue through the barn, level with the floor, its upper surface being covered with iron plate or tiles. By means of a small quantity of fuel a barn-full of corn in sheaves, properly disposed, might be dried in a few hours. The artificial method of drying grass here suggested will of course be unnecessary when the grass can be deprived of its water by the heat of the sun with sufficient rapidity, and without being exposed to the drenching influence of the rain of our northern climate. That rapid drying can be effected, even in wet seasons, in Scotland, I have had an opportunity of witnessing, in the case of an excellent sample of hay prepared during the summer of 1845, on the grounds of Mr. Fleming, of Barochan, for a specimen of which I am indebted to that gentleman. The only complaint which I have heard offered to the English plan of haymaking is the additional amount of labor required, but surely any rational excess of labor is preferable to the complete deterioration of the hay crop.

The constituents of the rye-grass, washed out by rain, would be principally the sugar and soluble salts.

The nature of the inorganic salts, both of the stem of the grass, when dried, as hay, and of the seeds, is as represented in the following tables.

100 parts of the stem and seeds were composed as follows :—

	Stem.	Stem.	Seed.
Water	15·50	19·00	11·970
Organic Matter -	79·52	75·72	82·548
Ash -	4·98	4·98	6·070

Table of Saline Matter in Stem and Seeds of Lolium perenne, (Rye-grass.)

	Stem.	Seed.
Silica -	64·57	43·28
Phosphoric Acid -	12·51	16·89
Sulphuric Acid -		3·12
Chlorine -	-	trace
Carbonic Acid -		3·61
Magnesia -	4·01	5·31
Lime -	6·50	18·55
Peroxide of Iron -	0·36	2·10
Potash -	8·03	5·80
Soda -	2·17	1·38

There is no doubt, from numerous other analyses which I have made, that these numbers undergo very considerable modifications on different soils.

A comparison of the two columns of this table adds another argument to that already brought forward against the practice of allowing rye-grass to come to seed before cutting it for hay, since the seed tends to remove a larger portion of phosphoric acid from the soil than the stem; the quantity of acid found in the

seed exceeding that in the stem by one fourth. A similar observation, with greater force, applies to the lime,
as the amount of this earth is two thirds greater in the
seed than in the stem. The quantity of alkalies is
twice as great in the stem as in the seed, while the
total ash of the seed is a sixth part superior in amount
to that of the stem.

CHAPTER VI.

ON BARLEY AND MALT DIET.

BARLEY AND MALT, WHEN NOT CRUSHED, ALTHOUGH STEEPED IN HOT WATER, ARE IMPERFECTLY DIGESTED BY COWS.—TOO LARGE A QUANTITY OF GRAIN DIMINISHES THE AMOUNT OF MILK.—BARLEY PRODUCES A GREATER QUANTITY OF MILK AND BUTTER THAN MALT.—DIFFERENCE IN THE ULTIMATE COMPOSITION OF BARLEY AND MALT.—DIFFERENCE IN THE AMOUNT OF NITROGEN IN BARLEY AND MALT.—DIFFERENCE IN THE SALINE CONSTITUENTS OF BARLEY AND MALT.—EFFECT OF THE PROCESS OF MALTING.

ALTHOUGH it might appear that the most correct method of determining experimentally the comparative nutritive effect of food would be to accustom an animal to a diet of one species of food, and then to substitute for a certain portion of it a definite quantity of that whose nutritive power was intended to be tried, and, lastly, to calculate the results, experience leads us to a different method of investigation. Physiology tends to show us, that an animal performing certain functions consumes an amount of oxygen daily, varying according to the state of the atmosphere and to other physical causes which are not always capable of appreciation. We adduce at once, then, from these circumstances, apart from experiments, that an animal consumes every day a different amount of fodder, and that, if it is not permitted to use as much food as shall repair the waste of its system, it must lose flesh and strength; and hence experiments made without a due attention to the physiological state of the animal must lead to conclusions which are not legitimate. The force of this observation we have had sufficient opportunities of observing, not only on the present but on other occasions, and it may be illustrated by the following example :— A cow, if fed for two days on an insufficient quantity of food, as indicated by loss of weight and diminution of milk, will require at least double that time to reach the condition from which it had deteriorated ; and the reason of this is obvious, because the partial starvation has caused it to lose a portion of the substance of its body, which requires a longer time to re-establish than to pull down. This rule is applicable to the dietary of men as well as the inferior animals. An increase of labor should always be accompanied with an increase of food, both at sea and in prison ; a short walk to one confined in a solitary cell calls for some augmentation of food. A slight increase of temperature, or the irritating influence of insects, will effectually diminish the milk of a cow, and indicates the propriety of increasing the amount of fodder. The first two of the following experiments demonstrates these positions in a striking manner. With the entire malt and barley the amount of grass was limited, but afterwards the hay was supplied *ad libitum.*

Experiment II.—Barley (entire) soaked in Boiling Water.

EXPERIMENT II.—BARLEY ENTIRE.

BROWN COW.

Days.	Date.	Milk.			Food.		Dung.			Weight of Cow.	Butter.			Temp.
	1845:	lbs.	oz.	drs.	Barley. lbs.	Grass. lbs.	lbs.	oz.	drs.	lbs.	lbs.	oz.	drs.	
1	June 24	22	11	12	2½	90	84	4	12	986				54°
2	— 25	21	12	15	2½	90	85	11	14	…				59
3	— 26	23	4	10	2½	90	96	4	14	…	3	13	9½	56
4	— 27	21	12	15	2¾	90	72	9	12	…				55
5	— 28	22	2	7	2½	90	80	5	14	…				55
6	— 29	20	12	6	2½	90	77	12	11	1009½				56
7	— 30	20	15	0	2¾	70	78	7	15	…				60
8	July 1	19	1	12	7½	70	71	8	12	…				53
9	— 2	18	8	10	7½	80	61	4	2	…	3	3	10	58
10	— 3	17	15	5	7½	80	61	9	15	…				51
11	— 4	17	7	1	2½	100	65	0	13	979				56
		226	8	13	47½	940	835	1	6	Loss, 7½	7	1	3⅕	

EXPERIMENT II.—BARLEY ENTIRE.

WHITE COW.

Days.	Date.	Milk.	Food. Barley.	Food. Grass.	Dung.	Weight of Cow.	Butter.	Temp.
	1845:	lbs. oz. drs.	lbs.	lbs.	lbs. oz. drs.	lbs.	lbs. oz. drs.	
1	June 24	21 15 1	2½	90	82 11 15	1044		
2	—— 25	21 3 4	2½	90	87 7 14	...		
3	—— 26	21 13 11	2½	90	86 10 2	...	2 8 0	
4	—— 27	21 2 9	2½	90	97 1 7	...		
5	—— 28	20 1 14	2½	90	83 1 15	1013½		
6	—— 29	18 14 15	2½	90	87 7 3	...		
7	—— 30	18 7 1	7½	70	83 12 8	...		
8	July 1	19 1 12	7½	70	83 6 0	...	2 10 0	
9	—— 2	18 13 5	7½	80	60 10 5	...		
10	—— 3	18 11 13	7½	80	68 1 11	...		
11	—— 4	19 9 14	2½	100	66 12 3	1010		
		219 15 3	47½	940	887 3 3	Loss, 34	5 2 0	

The result of this and the following experiment demonstrates the importance of reducing the food to a fine state of division.

Previous to this experiment, as will be observed by consulting the table of experiments on the effect of grass in feeding the cows, the animals were both gaining weight. By calculating the value of the barley as a nutritious body from the nitrogen contained in it, it was found that 2½ lbs. of barley contain as much albuminous nutriment as 10 lbs. of grass. The result of the experiment, however, shows that although this fact may be correct, yet that the conditions of the trial were not such as to prevent the animals from falling off both in milk and in weight. The true reason of the failure seems to have been, that the digestion of the barley was in some degree prevented by the want of power in the animal organs to rupture the husk of the grain. The result of the experiment demonstrates the importance of a certain amount of cookery in feeding cattle which are possessed of teeth only in one jaw.

The data which have served as the basis of the preceding calculations are included in the following table, as derived from repeated experiments :—

Water and Solid Matter in Food.

	Milk.	Dung.	Grass.	Barley.
Solid Matter - -	12·6	13·46	31·	90·54
Water - - -	87·4	86·54	69·	9·46

The white cow's milk on the second of July, or ninth day of the experiment, possessed the following composition, the specific gravity being 1,032 :

Water - - - - -	87·40
Soluble salts - -	0·17
Insoluble salts - ∤	0·42
Butter ⎞	
Sugar ⎬	12·01
Casein ⎠	

In several determinations the water in the milk of both cows was never found to vary more than a few tenths when properly dried.

In comparing this experiment with the preceding, by examining the proximate tables, (Table I. Appendix,) we find that while 100 lbs. of dry grass produce about 11¼ lbs. of dry milk, 100 lbs. of dry grass and entire barley mixed produce 8½ lbs. of dry milk. Grass alone produces a larger quantity of dung than mixed barley and grass fodder ; 100 lbs. of grass leaving 33½ lbs. of dung, while barley and grass produce only 30 lbs. of dung ; but 100 lbs. of the grass consumed, that is, the grass taken into the circulation of the animal, and not rejected in the form of dung, produces 17½ lbs. of dry milk, while 100 lbs. of the mixed barley and grass diet form only 12 lbs. of dry milk. This may proceed from the circumstance that more solid matter was actually contained in the grass than in the equivalent of barley employed ; but the cause becomes not so obvious when we consider that a portion of the barley was rejected entire along with the dung. The more probable explanation of the apparent anomaly may be, that the dung varies slightly in its composition ; the small difference of 3½ lbs. may be owing to this source of error in the calculation. Another important deduction from these two experiments in reference to economy is, that the total quantity of matter taken into the circulation daily is less, when grass is alone used, than when a mixed diet is employed; the daily consumption being of dry grass, by both cows, 33½ lbs., and of the mixed diet 42 lbs., being a difference of 9 lbs., or 4½ lbs. by each cow.

This fact may be explained by the circumstance, that there is a greater difficulty in digesting the grass, from its greater bulk, than in absorbing the constituents of the steeped barley, a large portion of which is in solution before being introduced into the stomach, and may be partially employed with greater rapidity in the process of producing heat, and partially be expelled as a liquid excretion.

Ultimate Analysis of the Experiment.—The ultimate composition of barley was found to be as follows :—

	I.		II.	III.	IV.
Carbon	46·11	41·64			
Hydrogen -	6·65	6·02			
Nitrogen	1·91	1·81	2·01	1·98	1·95
Oxygen	42·24	38·28			
Ash	3·09	2·79			
Water	–	9·46			
	100·	100·			

1st, 8·87 grains of barley, dried at 212°, gave, by combustion with chromate of lead, 15·04 carbonic acid, and 5·3 water.

2d, 14 grains gave, with lime and soda, 1·88 platinum=1·91 per cent. nitrogen.

3d, 0·923 gramme gave 0·288 gramme platino sal ammoniac=1·98 per cent. nitrogen.

4th, 0·834 gramme gave 0·262 platinum salt=1·95. nitrogen per cent.*

---

* For these two experiments I am indebted to Dr. Böttinger.

8

5th, 11·13 gave 1·57 platinum$=$2·01 per cent. nitrogen.

Calculating from the composition of the grass and barley, we find that the two cows consumed 304¼ lbs. of carbon during the course of the experiment, with a proportionate amount of the other ultimate ingredients. In this experiment it was observed, that some of the grains of barley were ejected from the intestines 24, 48, and even 72 hours after being swallowed, in an entire state, so that they must have been detained in some portion of the alimentary canal during that lengthened period without having undergone any appearance of digestion.

Experiment III.—Entire Malt soaked in Boiling Water.

EXPERIMENT III.—MALT ENTIRE.

BROWN COW.

Days.	Date.	Milk. lbs. oz. drs.	Food. Malt. lbs.	Food. Grass. lbs.	Dung. lbs. oz. drs.	Weight of Cow. lbs.	Butter. lbs. oz. drs.	Temp.
1	1845: July 9	18 9 15	2	90	78 8 8	994		60
2	— 10	19 6 11	3	90	80 9 14	...		61
3	— 11	19 7 5	3	90	87 2 10	...	2 14 2	57
4	— 12	19 11 0	3	90	68 15 8	...		57
5	— 13	18 15 14	3	90	70 0 0	985		55
6	— 14	18 10 6	6	80	65 11 4	...		59
7	— 15	18 14 5	9	80	81 7 9	...	3 7 8	60
8	— 16	18 10 10	9	80	68 7 8	...		57
9	— 17	19 3 7	9	80	71 8 4	...		56
10	— 18	19 9 2	9	80	77 9 2	1004		60
		191 2 11	56	850	750 0 3	Gain, 10	6 6 4	

EXPERIMENT III.—MALT ENTIRE.

WHITE COW.

Days.	Date.	Milk.			Food.		Dung			Weight of Cow.	Butter.			Temp.
		lbs.	oz.	drs.	Malt. lbs.	Grass. lbs.	lbs.	oz.	drs.	lbs.	lbs.	oz.	drs.	
1	1845: July 9	18	2	9	2	90	85	0	8	1050½				
2	— 10	19	1	0	3	90	98	0	4	...				
3	— 11	19	14	12	3	90	87	3	2	...	3	8	13	
4	— 12	20	8	5	3	90	80	12	10	...				
5	— 13	20	4	12	3	90	76	2	8	...				
6	— 14	20	9	9	6	80	80	1	2	1033				
7	— 15	21	8	14	9	80	87	b	15	...				
8	— 16	19	10	10	9	80	86	1	0	...	3	3	3	
9	— 17	20	10	14	9	80	83	6	12	...				
10	— 18	22	5	3	9	80	79	9	10	1018½				
		202	12	8	56	850	843	9	7	Loss, 32	6	12	0	

The malt was covered with boiling-hot water, and allowed to remain for twelve hours, in the first part of the experiment; in the latter period of the trial the malt was weighed out in three portions; the last portion was therefore subjected to a digestion of twenty-four hours. The mash water was always acid, and yet was relished by the cattle. This is opposed to the observation of some, who affirm, that acid liquors are not liked by cattle, although they are well known to be a luxury to pigs.

In consequence of the cattle having fallen off during the time in which they were fed with barley, farinaceous food was entirely discontinued, and a larger quantity of grass was substituted previous to the commencement of the experiment with malt. The result of this experiment is at once observed by an inspection of the table. The brown cow fell off in the amount of butter during the first five days, but increased during the remainder of the trial. The white cow gave a larger quantity of butter with malt than with barley. The milk of both cows increased very considerably, while the weight of the brown cow, which had decreased with the barley experiment, began to increase under the influence of the malt. We may infer, from the results of this experiment, the advantage of having a large portion of the food readily soluble and administered into the stomach of animals in this condition. The amount of butter would appear to depend more upon this provision than upon the quantity of matter soluble in ether existing in the food.

The mean of several dryings gave the composition of the dung,—water 86, solids 14. 3840 grs. of malt bruised gave 52·7 grs. of oil$=1·37$ per cent.

8*

According to the preceding trials, it appears that the
barley and malt experiments may be compared as follows :—(See Appendix I.)

    I. Milk :
        100 lbs. of hay and barley produce , 8·41 lbs. dry milk.
        100 lbs. of hay and malt produce   7·08   ditto.
   II. Butter :
        100 lbs. hay and barley produce   1·82 lbs. butter.
        100 lbs. hay and malt produce   2·07   ditto.

			Loss.
III. Weight of cattle :		lbs.	lbs.
Weight of cattle before barley experiment	-	2030	
Weight of cattle after ditto		1989	41
" " before malt ditto		2044	
" " after ditto	-	2022	22

It is obvious from this experiment that barley produced more milk than malt, even although it was only
partially digested ; that malt produced a little more
butter ; and that the cattle diminished in weight in both
experiments : most in the barley experiment, in consequence of a considerable quantity of it being thrown out
without being used by the system.

It is interesting to observe, that although the barley
and grass contained the largest amount of oil and wax,
they produced a smaller proportion of butter than the
malt and grass. This, however, may have-been in part
owing to the imperfect extraction of the solid ingredients in the barley experiments in consequence of the
husks remaining entire. The experiment is one, however, from which no deductions, to be entirely depended
on, are to be made. It demonstrates the necessity of
cooking barley, more especially when it is employed to

feed cattle. (1) 8·96 grains of malt, dried at the temperature of 212°, gave, when burned with chromate of lead, 14·3 carbonic acid and 5·66 water. (2) 7·86 grains gave 12·91 carbonic acid, and 5·01 water. This corresponds with, per cent:—

	I.	II.	III.	IV.	
Carbon -	43·93	44·780	–	–	·42·44
Hydrogen -	7·00	7·000	–	–	6·64
Nitrogen -	1·50	1·620	1·19	1·26	1·11
Oxygen	46·30	44·763	–	–	43·08
Ash -	1·27	1·777	–	–	1·68
Water	–	–	–	–	5·05
	100·	100·			100·

Total amount of constituents of food and dung, of both cows, in ten days:—

	Food.	Dung.	Consumption.	Each per Day.
	lbs.	lbs.	lbs.	lbs.
Carbon -	238·	102·	136·	6·80
Hydrogen -	32·2	12·43	19·77	0·99
Nitrogen -	9·06	4·	5·06	0·25
Oxygen - -	214·88	82·57	132·31	6·11
Ash -	34·22	21·80	12·42	0·62
				14·77

Experiment IV.—Crushed Barley steeped in Boiling Water.

As it appears from the preceding experiments that, when barley was given in an entire state, a considerable portion of the grain escaped the action of the digestive organs, in consequence of the interposition of the husk, it was necessary to try the effect of the grain as an article of food after it had been mechanically bruised.

EXPERIMENT IV.—BARLEY CRUSHED.

BROWN COW.

Days.	Date.	Milk. lbs.	oz.	drs.	Food. Malt, Hay. lbs.	Barley. lbs.	Grass. lbs.	Dung. lbs.	oz.	drs.	Weight of Cow. lbs.	Butter. lbs.	oz.	drs.	Temp.
1	1845: July 19	22	15	9	3	6	80	83	8	0	1004				63°
2	— 20	21	12	5	—	9	80	84	4	0					59
3	— 21	20	5	6	—	9	80	78	5	12					63
4	— 22	21	13	3	Hay. 30	9		86	4	8					58
5	— 23	19	14	6	35	9		69	4	2		3	10	3	61
6	— 24	19	5	8	35	9		82	11	15					58
7	— 25	20	12	2	35	9		87	11	15					60
8	— 26	21	1	3	13	9	26	100	5	11					57
9	— 27	22	1	5	35	9		87	3	15	984				61
10	— 28	21	11	2	35	9		85	5	15		3	15	4	58
11	— 29	22	5	13	28¾	9		80	10	9					58
12	— 30	22	3	15	30	9		83	14	12					58
13	— 31	22	7	4	21	9		80	3	15	1016½				59
14	August 1	22	1	0	30	9		55	7	0					
15	— 2	21	5	6	30	9		69	14	0		3	15	6	
16	— 3	20	10	9	30	9		87	5	15	1036				
		341	13	9	387¾ Hay. 3 Malt.	141	266	1310	10	0	Gain 32	11	8	13	

EXPERIMENT IV.—BARLEY CRUSHED.

WHITE COW.

Days.	Date	Milk lbs. oz. drs.	Malt, Hay. lbs.	Barley. lbs.	Grass. lbs.	Dung. lbs. oz. drs.	Weight of Cow. lbs.	Butter. lbs. oz. drs.	Temp.
1	1845: July 19	22 0 9	3	6	80	84 1 6	1018½		
2	— 20	22 6 14	—	9	80	80 10 15	...	3 3 4	
3	— 21	15 1 7 { Accident. }	— Hey.	9	80	74 10 0	1068		
4	— 22	22 4 0	30	9	—	81 15 9	...		
5	— 23	20 13 12	35	9	—	82 14 7	...		
6	— 24	21 3 11	35	9	—	87 4 2	...		
7	— 25	21 7 3	35	9	—	87 2 4	1020		
8	— 26	21 14 8	13	9	26	91 1 7	...	3 15 4	
9	— 27	22 9 14	35	9	—	87 6 15	...		
10	— 28	22 1 15	30	9	—	87 12 3	...		
11	— 29	22 10 9	27½	9	—	85 8 8	1082		
12	— 30	22 8 1	30	9	—	72 2 8	...		
13	— 31	22 8 13	25	9	—	77 9 4	...	3 6 4	
14	August 1	22 12 13	30	9	—	81 7 15	...		
15	— 2	21 2 10	30	9	—	88 3 11	...		
16	— 3	21 6 8	30	9	—	89 8 12	1075½		
		344 1 0 *7 0 0	375½ Hay.	141	266	1333 7 6	Gain, °57	10 8 12	
		351 1 0	3 Malt.						

* The 7 lbs. of milk added is that supposed to be lost by the accident, July 21.

Experiment V.—Crushed Malt digested in Boiling Water.

This experiment was intended as a parallel experiment with the preceding trial with crushed barley.

EXPERIMENT V.—MALT CRUSHED.

BROWN COW.

Days.	Date.	Milk.			Food.			Dung.			Weight of Cow.	Butter.			Temp.
					Malt.	Barley.	Hay.								
	1845:	lbs.	oz.	drs.	lbs.	lbs.	lbs.	lbs.	oz.	drs.	lbs.	lbs.	oz.	drs.	
1	August 14	20	12	10	6	3	26¼	89	14	0	1010				59°
2	— 15	19	2	2	9	—	26¼	78	12	0	...				56
3	— 16	19	10	0	9	—	26¼	55	8	0	...	3	7	5¼	57
4	— 17	18	11	1	9	—	27	84	10	15	...				53
5	— 18	18	13	7	9	—	30	78	0	8	...				57
6	— 19	18	13	12	9	—	24¼	84	12	2	998½				58
7	— 20	18	10	8	9	—	30	64	1	15	...				55
8	— 21	19	10	3	12	—	30	76	14	12	...				56
9	— 22	19	10	9	12	—	20	72	9	8	...	4	3	5	58
10	— 23	19	4	8	12	—	25	78	0	8	...				59
11	— 24	19	8	8	12	—	21½	85	3	0	...				59
12	— 25	20	6	5	12	—	17¼	78	10	1	...				59
13	— 26	19	11	15	12	—	26	77	4	9	996½				59
14	— 27	19	4	3	12	—	27¾	71	0	7	...	3	4	1	59
15	— 28	19	4	0	12	—	26½	83	7	0	...				61
16	— 29	18	8	13	12	—	26	79	0	12	1023½				63
		309	14	8	168	3	409¾	1237	14	1	Gain, 13½	10	14	11¼	

EXPERIMENT V.—MALT CRUSHED.

WHITE COW.

Days.	Date.	Milk.			Food.			Dung.			Weigh of Cow.	Butter.			Temp.
	1845:	lbs.	oz.	drs.	Malt. lbs.	Barley. lbs.	Hay. lbs.	lbs.	oz.	drs.	lbs.	lbs.	oz.	drs.	
1	August 14	22	0	3	6	3	30	94	3	8	1052				
2	— 15	21	14	15	9		30	75	11	0	...				
3	— 16	21	8	2	9		30	88	1	0	...	3	2	3¾	
4	— 17	20	6	15	9		30	54	0	8	...				
5	— 18	20	11	9	9		30	84	10	0	...				
6	— 19	20	11	3	9		27	76	11	0	...				
7	— 20	20	3	3	9		30	79	11	15	...				
8	— 21	21	14	14	12		30	77	9	12	...	3	10	9	
9	— 22	22	3	8	12		30	79	15	8	...				
10	— 23	22	7	0	12		36½	82	5	15	...				
11	— 24	21	6	9	12		30	79	0	10	...				
12	— 25	23	1	4	12		28	84	3	6	1028				
13	— 26	23	0	11	12		26½	74	3	15	...	2	14	11	
14	— 27	22	3	7	12		28	77	5	13	...				
15	— 28	21	10	1	12		28	74	15	4	1046				
16	— 29	19	10	12	12		28¼	76	13	13					
		345	1	8	168	3	472½	1259	10	15	Loss, 6	9	11	7¾	

EXPERIMENT VI.—EXCESS OF BARLEY CRUSHED.

BROWN COW.

Days.	Date.	Milk. lbs. oz. drs.	Food. Barley. lbs.	Hay. lbs.	Dung. lbs. oz. drs.	Weight of Cow. lbs.	Butter. lbs. oz. drs.	Temp.
1	1845: August 30	18 15 0	12	22¼	87 2 14	1023¼		65°
2	— 31	18 5 3	12	23	78 4 11	...	3 4 2	62
3	September 1	19 1 14	12	30	75 13 0	...		63
4	— 2	20 4 9	12	27	82 11 13	...		62
5	— 3	19 1 11	12	30	80 12 0	992		57
		95 12 5	60	132¼	404 12 5	Loss, 31¼	3 4 2	

WHITE COW.

Days.	Date.	Milk. lbs. oz. drs.	Food. Barley. lbs.	Hay. lbs.	Dung. lbs. oz. drs.	Weight of Cow. lbs.	Butter. lbs. oz. drs.	Temp.
1	1845: August 30	20 11 3	12	24¼	76 1 12	1046		
2	— 31	21 0 8	12	24	77 8 4	...	2 13 3¼	
3	September 1	21 14 7	12	30	75 12 10	...		
4	— 2	22 3 11	12	23	80 2 6	...		
5	— 3	21 2 8	8	30	68 13 2	1056		
		107 0 5	56	131¼	378 6 2	Gain, 10	2 13 3¼	

Experiment VI.—Larger Quantity of Crushed Barley steeped in Boiling-Water.

In the preceding malt experiment the amount of grain was pushed farther than in the case of barley; it was therefore considered advisable to give a similar trial to that grain. The result shows that no advantage is gained by the administration of so much grain, and that a deteriorating effect is induced. The cause of this seems to depend on the excess of nutritive over calorifient food, as will be afterwards explained.

Comparison of Experiments IV., V., and VI.

I. Milk.

> 100 lbs. of mixed barley, hay, and grass produced 8·17 lbs. milk. (Appendix I.)
>
> 100 lbs. of mixed malt and hay produced 7·95 lbs. milk.

II. Butter.

> 100 lbs. barley, hay, and grass produced 1·95 butter.
>
> 100 lbs. malt and hay produced 1·92 butter.

III. Weight of cattle.

	lbs.	Gain.	Loss.
Weight of cattle before barley experiment	2022		
— after — —	2111	89	
— after malt —	2069	...	42

According to this view of the experiment, it appears that the malt produces a smaller amount of milk and butter when combined with hay than in the barley experiment, and that the cattle were losing weight, and consequent strength, daily; while with barley they were gaining weight daily. In whatever manner therefore, we view the experiment, this is an insurmountable objection to the use of malt,—that it is not capable when used in any quantity, comparatively with barley, to sustain the weight and consequent strength of animals. But there is another aspect in which the experiment should be examined, and this is obviously the correct one, since a larger quantity of malt was used than of barley. If we consider the hay a constant quantity, and then calculate the amount of product which would comparatively result from each grain, the consequences would be as follows, (Appendix I. :)—

I. Milk.

    100 lbs. of barley would produce by Experiment IV. 34·6 lbs. dry milk.

    100 lbs. of malt would produce by Experiment V. 26·2 lbs. dry milk.

II. Butter.

    100 lbs. of barley would produce by Experiment IV. 7·66 lbs. butter.

    100 lbs. of malt would produce by Experiment V. 6·35 lbs. butter.

By the present mode of comparison then it appears that, in every point of view, malt is inferior to barley as an article of diet for cattle, as it gives less milk and butter, and diminishes the live weight, instead of increasing it, which barley does under the same circumstances.

All these practical results are explained by the chemical examination of the barley and malt, which will be subsequently stated and discussed. In the mean time it may be sufficient to intimate that the deductions now made from the practical trials are in exact accordance with experiments conducted in the laboratory. The soluble salts are much diminished in the malt, and hence a larger quantity of the grain would be required than of barley to produce the salts of a given amount of milk. The quantity of nitrogen is inferior to that in the barley, and hence malt must be inferior in nutritive agency to the barley, in comparing equal weights, while the quantity of sugar being greater, the amount of butter produced might be equal or nearly so to that formed from barley, as is observable in some of the experiments.

On the Chemical Nature of Barley and Malt.

From the nature of malting it might be expected that a considerable difference would exist between barley, before and after being subjected to this process.

In the following experiment the malt was made from the same specimen of barley, so as to enable a tolerably correct comparison to be instituted.

I. *Difference in ultimate Composition.*—The barley, when subjected to organic analysis with chromate of lead, was found to possess the following composition :—

		I.	II.	III.	IV.
Carbon	41·64	46·11			
Hydrogen	6·02	6·65			
Nitrogen	1·81	2·01	1·91	1·98	1·95
Oxygen	37·66	41·06			
Ash	3·41	4·17	4·30	3·27	
Water	9·46				
	100·	100·			

The first column exhibits the composition of the bar

ley in its natural state ; the second represents the constituents of the barley when dried at the temperature
of 212°.

Malt from the same barley was also analyzed, and
the following result obtained :—

		I.	II.	III.	IV.
Carbon	42·44	43·930	44·78		
Hydrogen	6·64	7·000	7·06		
Nitrogen	1·11	1·290	1·26	1·504	1·62
Oxygen	43·08	46·510	45·13		
Ash	1·68	1·270	1·77		
Water	5·05				
	100·	100·	100·		

In the first column we have the composition of malt
in its natural state, and in the other columns its constituents at 212°, as determined by two analyses, the
first column being calculated from the third column or
second analysis, founded upon the determination of the
amount of loss sustained when the grain was subjected
for some days to the heat of boiling water in a water
bath. If we now divide the constituents of barley and
of malt by their equivalents, or combining proportions,
we shall be able to form some idea of the change which
has taken place in the barley during its conversion
into malt. The following is the result :—

	C.	H.	N.	O.
Barley	123	106	2	82
Malt	119	112	0	90
Difference $\{$	4	0	2	0 loss
	0	6	0	8 gain.

If we consider that 100 parts by weight of barley are
converted by the process of malting into eighty parts

by weight of malt, we shall have the following formulæ :—

				C.	H.	N.	O.
Barley	-	-	-	123	106	2	82
Malt		-		90	85	0	69

| | | | | 33 | 21 | 2 | 13 loss; |

and the barley and its equivalent amount of malt will
then stand as follows, per cent., and in eighty parts :—

	Barley.	Malt.
Carbon -	41·64	33·95
Hydrogen - -	6·02	5·31
Nitrogen - - -	1·81	0·88
Oxygen	37·66	34·46
Ash =	3·41	1·34
Water - -	9·46	4·06
	100·	80·

Hence it appears that four equivalents of carbon have
disappeared in the malting, without doubt in the form
of carbonic acid, and an equivalent of nitrogen has also
been removed in the shape of albumen, possibly in part
as ammonia, while the malt contains six of hydrogen
and eight of oxygen in excess over that contained in the
barley. The odd atoms of oxygen are probably an
error of experiment; and if we allow this then, we shall
have a difference in the malt, in the fact of six equivalents of water (6 H. 6 O.) having been added to it during
the malting process; and this admits of explanation
from the circumstance, that one of the important alterations in malting consists of the conversion of starch into
sugar. Now the difference between starch and sugar
is simply that the latter contains more water than the

9*

former, the composition and difference of these substances being as follows :— '

	c.	h.	o.
Starch	12	10	10
Sugar	12	12	12
	0	2	2 difference.

II. *Difference in the Amount of Nitrogen, and consequent Nutritive Power of Malt and Barley.*—In the
preceding formulæ the quantity of nitrogen lost has been
somewhat exaggerated. In the formulæ for malt the
true amount of nitrogen approaches nearly $1\frac{1}{2}$ equivalent, or 1·4 ; but the quantity of nitrogen in different
parts of the same sample of malt varies very remarkably, indeed to such a degree that the results obtained
by three analysts, who had obtained almost identical
numbers for the nitrogen in barley, differed as much as
from 1·19 to 1·62. This indeed is a circumstance
which might be anticipated from the nature of the process of malting, and is one which renders malt a very
objectionable substance as an article of nourishment,
since, in the same specimen, different portions would
vary so much according to the preceding data, as that
73 lbs. of one part would produce as much effect in
the nourishment of an animal as 100 lbs. of another
portion.

If we estimate the albuminous principles of grain to
contain 16 per cent. of nitrogen, then the amount of
these substances in the barley examined will amount
to 12·56 per cent., while the percentage of these principles in the malt will only be, by the lowest estimate
of nitrogen, 7·43, and by the highest result it will be
10. So that the relative nutritive powers of barley and
malt, according to these estimates, will be as follows :

59 barley = 100 malt, according to lowest estimate.
79 — = 100 — highest —

These important facts render it also obvious that the
difference in the amount of carbon in the two analyses
of malt previously given may not have risen from errors
of analysis, but from a difference actually in the consti
tution of the malt.˙ That which contained the largest
amount of nitrogen would also contain the greatest
amount of carbon. Indeed it may be looked upon as
a rule with reference to nutritive bodies, generally
speaking, that their power of sustaining the animal
system depends, in relation to their ultimate composition, upon the amount of carbon and nitrogen which
they contain. Some have endeavored to prove that it
is the amount of carbon to which we are to look in de-
ciding upon the relative nutritive power of food, while
others have advocated the importance of nitrogen in
forming such estimates. It seems, however, certain,
from a careful study of all the facts, that such general
rules cannot safely be adopted, since, in the case of oils,
we have examples of substances containing much carbon which are yet incapable of supplying the waste of
the muscular substance of animals, and are therefore to
be excluded from the rank of true nutritive principles ;
while, again, we have gelatine or jelly containing nearly as much nitrogen as muscular fibre itself, which has
been proved to be incapable of supporting animal existence, in the manner in which we understand that expression when applied to beef or true muscular fibre.
Dogs, for example, have been made to live for months
on pure albuminous matter ; an experiment undoubtedly somewhat unnatural, and incapable of being persist-
ed in for any more considerable period. Again, the

true unsophisticated American Indians, near the sources
of the Missouri, during the winter months, are reported
to subsist entirely upon dried buffalo flesh—not the fat
portions, but the muscular part ; and during this period
those primitive inhabitants of the prairies, as they are
made up of nomade tribes, every man being at war
with his neighbor, are destitute of the means of supplying themselves with vegetable food, as they have no
gardens, nor any species of cultivation ; but, more particularly during their subsistence on dried *pemmican,*
they are described by travellers who are intimate with
their habits of life as never tasting even the most minute portions of any vegetable whatever, or partaking
of any other variety of food.   These facts, then, tend
to show that albuminous tissue is of itself capable of
sustaining life.   But we have no example of animals
being capable of subsisting on gelatine or glue ; on the
contrary, we have proof that animals, when restricted
to the use of this species of matter, become deteriorated
in health.   In the mean time, therefore, it may be advisable to admit, that we are unacquainted with the exact
position gelatine holds in the nutritive category, and to
place it among the exceptions to the nearly general fact,
that the amount of nitrogen is an important element in
calculating the value of a substance as a nutritive agent.
When we reflect that animals subsisting upon vegetable food contain an equal quantity of gelatine as a constituent of their tissues with those which have partaken
of animal food alone, we can scarcely fail to conclude
that gelatine, or glue, is a product of the alteration of
albuminous matter, and a stage in its downward pro-
gress to the state of urea, or an ammoniacal salt, for the
purpose of being removed from the system ; and hence,

that it is not capable of forming the muscular or highest order of animal matter. With this exception, then, we are inclined to adopt the idea, that the amount of carbon and nitrogen present in a substance supplies us with one of the data for calculating its capability to supply the waste of the muscular system of animals, the relation of the two substances, to constitute an efficient nutritive substance being nearly as 70 to 9 of their equivalents, represented by the formula 70 C. 9 N., the relation in gelatine being nearly as 66 C. $8\frac{3}{4}$ N. The first formula will be found useful for practical purposes ; since, when we have determined by analysis the amount of carbon and nitrogen consumed by an animal, we can distinguish, by dividing the respective numbers by those of the formulæ, how many equivalents of the total carbon are associated with the nitrogen, and employed by the animal for the purpose of supplying the waste of the muscular system, or by bearing in mind that the relation of nitrogen to the carbon of muscular fibre is as 16 to 53 nearly, we can discover the amount of carbon united to the nitrogen by the simple formula $\dfrac{53 \times a}{16}$.

In a cow, for example, consuming per day 7 lbs. of carbon and $\frac{1}{4}$ lb. of nitrogen, it will be found how insignificant is the quantity of carbon required for repairing the loss of the muscular system, $\dfrac{53 \times \cdot 25}{16} = 0 \cdot 828$ lbs. Hence we see that $6 \cdot 172$ lbs. of carbon of the daily food of a cow must be employed for a purpose totally distinct from proper nutrition. We are at present acquainted with only one other purpose for which the carbon of the food can be employed, viz. for the generation of animal heat throughout the body; a function undoubtedly carried on, not only in the lungs, but also throughout the entire capillary system of the skin, at least in man and perspiring animals. If this view be correct, then it follows that upwards of 6 lbs. of carbon are expended by a cow daily in the production of animal heat. And as 1 lb. of carbon, when combined with the necessary amount of oxygen to form carbonic acid, gives out as much heat as would melt 104·2 lbs. of ice, it is evident that the quantity of ice capable of being melted by the heat generated by a cow in one day would amount to upwards of 625 lbs., or it would heat 1 lb. of water 87,528°. It would consume at the same time the enormous quantity of 330429 cubic inches of oxygen, or 191¼ cubic feet of this gas ; and as this amounts to one-fifth of the atmospheric air, we find that a cow, consuming 6 lbs. of carbon for respiratory purposes, would require 956¼ cubic feet of atmospheric air, a sufficient indication of the immense importance of a free ventilation in cow-houses, and of the danger of overcrowding, if the animals are expected to retain a healthy condition. It is not to be supposed that the food, destined for the purposes of respiration, is thrown off in the form of carbonic acid as soon as it passes into the circulation. On the contrary, we may infer, from various experiments, that it remains for some time in the system in the condition of preparatory fuel, if we may so speak, undergoing during that period certain changes necessary for enabling it to take part in the respiratory function.

III. *Difference in the Saline Constituents of Barley and Malt.—Barley.*—The amount of inorganic matter existing in different specimens of barley varies very considerably. This might be anticipated from the fact, which is now generally admitted, that the azotized or nutritive principles of grain or seeds bear a relation to the phosphoric acid present. (*Liebig.*) Thus, if the quantity of phosphoric acid in barley be small, it will follow that the amount of nitrogen will be proportionally deficient, and that the nutritive effect of the grain will be comparatively low in the scale, because the solubility of the albuminous matters, and therefore their capability of being carried into plants, appears to depend on the presence of the phosphates. In the analyses which have been published of this nature, the experimenters have omitted to state whether the husks were included in the amount of grain burned by them ; in the following results the omission has been filled up. In the three last experiments, 1000 grains of the barley were burned ; in the first, the amount ignited was about fifty grains, but the ash was perfectly white, containing not a trace of charcoal.

Barley	Flour.			With husk.			
	I.	II.		III.	IV.	V.	VI.
Inorganic matter, per cent. -	4·17	3·87		3·27	3·20	3·02	2·70

In all these experiments the grain was dried at 212°, and each number represents the percentage of inorganic matter. The specimens were all different, but the fi t result was obtained from the barley used in the experiments. These numbers differ to a considerable degree from the experiments hitherto published. The following are such as have come in our way with reference to the per-centage amount of ash in barley :—

I.	II.
1·80	2·70
Saussure.	Koechlin.

The first of these specimens was probably derived from the neighborhood of Geneva, the second was from Neufchatel, near the lake of that name in Switzerland.

The following was found to be the per-centage composition of the ash of barley :—

Silica	-		29·67
Phosphoric acid	-	-	36·80
Sulphuric acid	-	-	0·16
Chlorine	-		0·15
Peroxide of iron			0·83
Lime		-	3·23
Magnesia	-		4·30
Potash	-	-	16·00
Soda			8·86

Some chemists have found no alumina in the ashes of grain. Boussingault states that he generally finds traces, and in this respect our observations agree, and in some instances the quantity has appeared almost too considerable to be accidental.

Malt.—We are now in a condition to compare the influence of malting on the saline constitution of the barley. In this respect the results of the present experiments corroborate those made upon the amount of nitrogen contained in various specimens of malt, for we find that the quantity of saline matter varies considerably, although not more than in different specimens of barley; but we are drawn to the conclusion, that a substance so unequal in its composition in reference to the proportion between the soluble and insoluble saline ingredients· is scarcely to be recommended as a food capable of producing a steady effect. The following experiments exhibit the amount of saline matter in different samples of malt contained in 100 parts of the grain dried at 212° :—

With husk.

I.	II.	III.	IV.
2·38	2·00	2·43	2·46

Table of the Saline Constituents of Malt.—The following table presents the results of careful analyses of the ashes of malt :—

	I.	II.	III.
Silica -	28·74	28·65	28·98
Phosphoric acid	35·34	33·18	34·65
Chlorine -	Trace	0·36	
Peroxide of iron -	1·59	1·94	1·72
Lime	3·89	5·13	3·62
Magnesia - -	9·82		
Potash - -	14·54	11·72	
Soda - -	6·08	4·90	

To determine the nature of the saline ingredients removed from barley in the malting process, it was necessary to examine the solid constituents of steep water. For this purpose several gallons of steep water were evaporated to dryness, and yielded about half its weight of organic matter, consisting of albumen and sugar, &c.

100 grains of the salt containing this organic matter, dried at 212°, afforded ·878 nitrogen, which is equivalent to 5·49 per cent, of albumen. The salts consisted of alkaline phosphates, carbonates, sulphates, and chlorides.

Effect of the Process of Malting.—These analyses afford some information in reference to the process of malting, and to the change which the barley undergoes by this operation. One of the most striking alterations produced in the barley, by its being steeped in cold water for forty hours and upwards, is to diminish its weight. Equal volumes or measures of barley and malt were found respectively to weigh 424 and 325 grains. This would give us 100 parts by weight of barley, equivalent to 76·65 of malt; but as barley expands slightly, or increases in bulk by steeping and conversion into malt, the difference between the two conditions is scarcely so considerable. In three returns obtained by us from maltsters, we are informed that—1st, 27 cwt. of barley become $22\frac{1}{2}$ of malt, or equivalent to 100 barley. and $83\frac{1}{3}$ malt; 2d, a bushel of barley weighing 55 lbs. becomes, when malted, from 43 to 45 lbs., or equal to 100 barley, and from 78·2 to 82 lbs. malt; 3d, a bushel of barley weighing 55 lbs. becomes 43 lbs. when malted, or as 100 to 78·2. The mean of all these indicates a loss which the barley sustains by malting of nineteen per cent., and upwards; or the loss might be taken approximately at twenty per cent., or one-fifth. The whole of this loss is not, however, solid matter; for, according to our trials, barley, when not crushed, contains 13·1 per cent. of water, and malt in the same condition 7·06 per cent. of water, capable of being dissipated at the temperature of 212°. Hence, of the nineteen per cent. of loss sustained by the barley in malting, six per cent. is water. There thus remain therefore only thirteen per cent. to be ascribed to solid loss. The quantity of saline matter

removed from the barley is considerable. A mean of
several trials gives, for the ash of barley, three per
cent., and for that of malt 2·52 per cent. Now as 100
barley are equal to 80 malt, the quantity of ash which
malt should contain is 2·42, if the loss of inorganic and
organic matter were equable, which we observe it to be
almost approximately from this experiment ; for the
relation of the ash which has disappeared, or 0·48 per
cent., bears almost the same proportion to the organic
matter also removed as the total quantity of ash in
barley does to the total organic matter of that grain.
Thus barley contains eighty-four per cent. of dry organic matter, and three per cent. of ash, while malt has
lost 0·48 per cent. of ash, and 12·52 of organic matter ;
and by calculation we have—

$$\text{As} \quad 3 : 0·48 .: 84 : 13·4;$$

a remarkable coincidence, as if proving that water is
incapable of removing the ash of plants until the organic matter has undergone such a change as to allow
the ash to separate. We have thus an argument in
favor of the subsistence of a chemical union between
the inorganic and organic matter of which the substance
of farinaceous grain is composed. Should this view
be well founded, the amount of ash in grain, we might
expect, would bear a constant ratio to the dry organic
matter by weight in whatever soil it might be grown.
It would also follow that cold water will not take up
saline matter from an entire seed simply by washing
or slight digestion.

The loss sustained by barley in malting may perhaps
be stated as follows :—

Water -	6·00
Saline matter	0·48
Organic matter -	12·52
	19·00

The nature of the saline matter removed from the barley is exhibited in the analysis of steep-water ash, although it is not so easy to explain the source of some of the constituents. We observe, in the first instance, that silica has been removed from the barley; the steep-water ash containing about 2 per cent. of silica. That this substance is united with potash is obvious from the gelatinization which occurs when hydrochloric acid is added to the steep salt. The origin of the carbonic acid, or rather its condition of union, is not so apparent: it might be attributed to the impurity of the water, but the presence of a minute amount only of lime opposes this explanation. The water used in the steep was the Clyde water, which contains chalk in solution, and sulphate of lime. To this source the sulphuric acid may owe its presence. The richness of the steep water in alkaline salts suggests its employment as a manure. A considerable part of the organic matter of the barley is dissipated in the form of carbonic acid, but a large portion of the albumen and sugar is also dissolved in the water, the solution of the albuminous matter being probably assisted by the action of the phosphates, which are capable of dissolving, it is well known, some of its forms, more particularly casein. The quantity of nitrogen obtained from the steep salt, when evaporated and dried at 212°, was very considerable, being equivalent to five and a half per cent. of albumen, if the whole of the nitrogenous matter existed in the form of that principle. But, besides this substance, there was present also a large quantity of other organic matter in the steep solution, since the steep salt, when dried at 212°, and then ignited, lost upwards of forty per cent. of its weight.

The views which we have been discussing of the difference in the chemical composition of barley and malt are sufficient to render it obvious that malt is a much more expensive substance, irrespective of duty, than barley for feeding, inasmuch as it is in reality barley deprived of a certain portion of its nutritive matter and salts. The only advantage which it seems to hold out in cattle feeding is the relish which it gives to a mash ; but as this depends entirely upon the sugar which it contains, and which has been produced from the starch of the barley, it is obvious that the same flavor may be imparted by the addition of an equivalent amount of molasses or sugar, should it be considered expedient. But we believe this mixture would be opposed to the true laws of dieting, to be subsequently discussed : we have always, however, found steeped barley to be highly relished by cattle. Malt, however, from the diastase it contains, has the power of speedily converting the starch of barley into sugar : according to Payen, a handful of malt would be sufficient to saccharize several pounds of barley in the steep.

10*

CHAPTER VII.

EFFECT OF MOLASSES, LINSEED, AND BEANS, IN THE PRODUCTION OF MILK AND BUTTER.

MOLASSES GIVES LESS MILK AND BUTTER THAN A DIET CONTAINING MORE NITROGEN.—LINSEED GAVE LESS BUTTER THAN BEAN MEAL, ALTHOUGH CONTAINING MORE OIL, PROBABLY IN CONSEQUENCE OF THE CONSTITUENTS OF BEANS BEING IN THE NATURAL PROPORTION TO RESTORE THE WASTE OF THE ANIMAL SYSTEM.

The following experiments were instituted for the purpose of determining the effect of other important species of food to serve as objects of comparison. The tables which follow include the result experienced by feeding both cattle on barley and molasses, barley and linseed, and on bean meal. The object of continuing the barley with the molasses and linseed was to enable an appreciation to be more readily formed of the effect of the substitution of one kind of food for another, without subjecting the animal to an entire change of diet. This mode of procedure was suggested by physiological principles, and was conducted in the same manner as the dieting of the human species. The experiments, however, have shown that attention to this point is not so indispensable as might at first sight appear, since a complete change of food is often followed by an increase of the secretions of milk and butter.

For steady and unwearied assistance in the whole of these experiments, I have been much indebted to my intelligent pupil, Mr. Hugh B. Tennent. Most of the weighings, &c. of food were made by us conjointly, and none of them without the presence of one or both of us.

EXPERIMENT VII.—BARLEY AND MOLASSES.

BROWN COW.

Days.	Date.	Milk. lbs. oz. drs.	Food. Barley. lbs.	Food. Molasses. lbs.	Food. Hay. lbs.	Dung. lbs. oz. drs.	Weight of Cow. lbs.	Butter. lbs. oz. drs.	Temp.
1	1845: August 4	18 8 1	9	0	25½	84 10 4	1036		60°
2	— 5	22 1 0	9	3	23	85 9 0	...		63
3	— 6	21 15 5	9	3	30	94 12 12	...	3 10 7	60
4	— 7	21 9 8	9	3	24½	93 9 15	...		60
5	— 8	21 1 7	9	3	30	97 13 4	...		57
6	— 9	20 6 2	9	3	30	79 9 0	...		59
7	— 10	19 13 8	9	3	26	77 15 2	...		59
8	— 11	19 7 13	9	3	25	70 6 6	...	3 10 7	59
9	— 12	19 9 5	9	3	30	85 4 0	...		59
10	— 13	19 3 6	9	3	26	81 14 4	1038		59
		203 11 7	90	27	269	851 7 15	Gain, 2	7 4 14	

EXPERIMENT VII.—BARLEY AND MOLASSES.

WHITE COW.

Days.	Date.	Milk.			Food.			Dung.			Weight of Cow.	Butter.			Temp.
		lbs.	oz.	drs.	Barley. lbs.	Mo-lasses. lbs.	Hay. lbs.	lbs.	oz.	drs.	lbs.	lbs.	oz.	drs.	
1	1845: August 4	20	9	8	9	0	23¼	87	0	0	1075½				
2	— 5	21	9	2	9	3	30	87	5	0	...				
3	— 6	23	0	2	9	3	30	98	13	12	...	3	4	5	
4	— 7	23	4	15	9	3	21	87	6	15	...				
5	— 8	23	7	6	9	3	30	84	4	0	...				
6	— 9	23	1	14	9	3	30	86	2	15	...				
7	— 10	22	5	13	9	3	26¼	80	12	2	...				
8	— 11	21	13	10	9	3	30	82	8	5	...	3	4	5	
9	— 12	22	10	2	9	3	30	86	6	8	...				
10	— 13	22	7	11	9	3	26	86	1	0	1106				
		203	12	11	90	27	274	866	12	9	Gain, 30½	6	8	10	

EXPERIMENT VIII.—BARLEY AND LINSEED.

BROWN COW.

Days.	Date.	Milk.			Food.			Dung.			Weight of Cow.	Butter.			Temp.
		lbs.	oz.	drs.	Barley. lbs.	Linseed. lbs.	Hay. lbs.	lbs.	oz.	drs.	lbs.	lbs.	oz.	drs.	
1	1845: Sept. 4	19	10	9	9	3	24¾	75	6	4	992				58°
2	— 5	20	6	3	9	3	30	77	7	0	...				56
3	— 6	20	0	14	9	3	26¼	81	10	3	...	3	11	8½	55
4	— 7	20	8	0	9	3	27½	82	8	2	...				55
5	— 8	20	9	5	9	3	21	84	8	8	992				57
6	— 9	20	8	14	9	3	30	86	11	4	...				59
7	— 10	20	5	6	8	4	26½	77	12	12	...	3	7	0	57
8	— 11	20	13	5	6	6	30	70	1	6	...				53
9	— 12	22	2	4	6	6	22	76	5	8	...				52
10	— 13	20	10	2	6	6	27½	76	10	4	1027				52
		205	10	4	80	40	267½	789	1	3	Gain, 35	7	2	8½	

EXPERIMENT VIII.—BARLEY AND LINSEED.

WHITE COW.

Days.	Date.	Milk.			Food.			Dung.			Weight of Cow.	Butter.			Temp.
	1845:	lbs.	oz.	drs.	Barley. lbs.	Linseed. lbs.	Hay. lbs.	lbs.	oz.	drs.	lbs.	lbs.	oz.	drs.	
1	Sept. 4	21	2	8	9	3	18¾	77	10	15	1056				
2	— 5	22	1	5	9	3	27¾	86	1	7	...				
3	— 6	22	4	0	9	3	25	87	7	11	...	3	6	8½	
4	— 7	23	11	3	9	3	25¾	87	7	11	1050⅕				
5	— 8	23	11	2	9	3	20¼	74	2	0	...				
6	— 9	23	12	3	9	3	30	83	7	3	...				
7	— 10	23	5	4	8	4	27¾	70	6	7	...	3	6	12	
8	— 11	23	6	13	6	6	26	70	6	7	...				
9	— 12	24	6	0	6	6	20	72	0	0	...				
10	— 13	21	13	2	6	6	28	76	5	12	1060				
		230	9	8	80	40	249¼	785	7	9	Gain, 4	6	13	4½	

Experiment IX.—Results from Feeding on Bean Meal.

BROWN COW.

Date.	Milk.	Food.			Duog.	Weight of Cow.	Butter.	Temp.
		Linseed.	Bean Meal.	Hay.				
	lbs. oz. drs.	lbs.	lbs.	lbs.	lbs. oz. drs.	lbs.	lbs. oz. drs.	
1845: September 14	19 3 2	4	8	30	74 5 10	1023		56
—— 15	20 8 4	...	12	30	70 3 8	...		46
—— 16	19 0 4	...	12	28	74 12 0	...	3 11 10	55
—— 17	19 10 6	...	12	28	84 3 4	...		55
—— 18	21 5 12	...	12	30	80 10 4	...		58
	99 11 10	4	56	119	384 2 10	...	3 11 10	

WHITE COW.

Date.	Milk.	Food.			Dung.	Weight of Cow.	Butter.	Temp.
		Linseed.	Bean Meal.	Hay.				
	lbs. oz. drs.	lbs.	lbs.	lbs.	lbs. oz. drs.	lbs.	lbs. oz. drs.	
1845: September 14	23 14 3	4	8	25	84 4 8	1060		
—— 15	22 5 6	...	12	24	74 0 0	...		
—— 16	21 2 6	...	12	20	64 8 0	...	3 12 6$\frac{1}{2}$	
—— 17	22 15 2	...	12	20$\frac{1}{2}$	60 1 12	...		
—— 18	25 4 14	...	12	30	72 15 0	...		
	115 9 15	4	56	119$\frac{1}{2}$	355 13 4	...	3 12 6$\frac{1}{2}$	

From the three preceding tables we learn the follow-ing particulars in reference to the milk and butter of the cows :—

I. Milk :   lbs.
 1000 lbs. of hay, barley, and molasses produce
  of dry milk - - 80·6
 1000 lbs. of hay, barley, and linseed - - 84·5
 1000 lbs. ditto bean meal - 81·3
II. Butter :
 1000 lbs. of hay, barley, and molasses produce
  butter - 21·9
 1000 lbs. of hay, barley, and linseed - 21·5
 1000 lbs. ditto bean meal - 22·5

or, considering the hay a constant quantity, then we have the results as follows :—

I. Milk :   lbs.
 1000 lbs. barley and molasses produce of milk 237
 1000 lbs. ditto linseed - 257
 1000 lbs. bean meal - - - - 252
II. Butter :
 1000 lbs. barley and molasses produce of butter 64·5
 1000 lbs. ditto linseed - 65·7
 1000 lbs. bean meal - - 70·0

By examining the 4th Table in Appendix, we ob-serve the comparative effect of linseed and beans, during equal periods, in producing milk and butter. In the case of the white cow, particularly, the results are quite unequivocal ; for while during five days the milk pro-duced by beans was equal to the mean of that produced by linseed during ten days, the amount of butter under the bean diet was greater than under any other kind of food whatever. This is an important fact in reference to the source of butter in the food, since the linseed meal, employed in the experiments, contained twice as much oil as the bean meal. In the brown cow also the quantity of butter was greater, especially during the second five days, with beans than with linseed. Molasses produced in the brown cow also a larger quantity of butter than the linseed, while the amount was slightly inferior to that produced by the beans. These facts, then, are not agreeable to the opinion that the amount of butter afforded by a cow is a test of the amount of oil contained in the food ; and hence we are not entitled to recommend oily food as preferable for the production of butter and of fat in animals to food which experience teaches us to be productive of this effect, although less rich in oleaginous matter. Indeed, the constant practice of giving oil cake to cattle is not an argument in favor of the importance of oil in the formation of fat, since from oil cake as much of the natural oil of the rape-seed or linseed has been removed by expression as mechanical means can effect. The oil-cake argument is so much the more, therefore, calculated to refute the objects to which it is generally applied.

The chemical composition of the linseed and bean meal is calculated to throw some light on the causes of the differences in the amount of products in the experiments. The following table represents the ultimate composition of linseed and bean meal, determined by combustion with chromate of lead.

Table of ultimate Composition of Linseed and Beans.

	Linseed.		Beans.	
		Dried at 212°.		Dried at 212°.
Carbon - -	42·51	49·55	40·76	45·59
Hydrogen - -	6·22	7·26		
Nitrogen	3·78	4·41	4·13	4·61
Oxygen	26·35	30·68		
Ash	6·94	8·10	3·22	3·96
Water - -	14·20		10·60	

Table of the Composition of the Ash of Linseed and Bean Meal. (Horse Bean.)

	Linseed.	Bean Meal.
Silica - - -	34·85	13·12
Phosphoric Acid -	25·22	35·26
Sulphuric Acid -	2·85	1·29
Chlorine - -	trace	1·75
Lime - - - -	6·95	5·18
Peroxide of Iron - - -	3·23	1·80
Magnesia - - - -	8·04	9·03
Potash - - - - - -	16·85	23·15
Soda - - - - - -	2·22	9·42

The great preponderance of alkaline salts in bean meal is observed distinctly in its incineration, as the ash fuses into a white salt, and, if care is not taken, will enclose charcoal, which can with difficulty be burned away. To avoid this obstacle the meal should at first be burned, with free exposure to air, at a low red heat.

From this and the preceding table we find that a given weight of bean ash contains a much larger quantity of phosphoric acid than the same amount of linseed; but as the ash of linseed is double in amount to that of the beans, there is present a larger per-centage of phosphoric acid than in beans.   Linseed, however, contains a large quantity of silica and sand, which is useless to the animal system.   The superior influence of beans in producing milk and butter is attributable to the constituents of milk existing with proper equilibrium.   They, therefore, restore the waste of the animal system in the proper proportions.

The present tables also seem to prove most conclusively, that the butter of the cows cannot possibly be produced from the wax and oil of the food, since the greater portion of the wax of the food reappears in the dung, (Table I. Appendix,) being expelled from the animal without change ; while the butter and wax of the dung greatly exceed all the oil and wax of the food. From these circumstances it is very much to be doubted, whether the wax of hay occupies any place in the production of the fat and butter of animals.   In all the experiments the wax of the dung was found always to vary slightly, so that it seems highly probable if the whole wax had been extracted from the dung, it would be found that all the wax of the food was excreted by the animals.

CHAPTER VIII.

QUANTITY OF MILK PRODUCED BY DIFFERENT KINDS OF FOOD.—EFFECT
OF GRASS IN PRODUCING MILK.—INFLUENCE OF VARIETY OF FOOD ON
MILK AND ON MAN.—ECONOMICAL DISHES FOR THE POOR.—EFFECT OF
BARLEY AND MALT ON MILK.—EFFECT OF MOLASSES, LINSEED, AND
BEANS ON THE PRODUCTION OF MILK.—INFLUENCE OF QUANTITY OF
GRAIN IN THE PRODUCTION OF MILK.—RATE AT WHICH FOOD IS
CHANGED INTO MILK.—RELATIVE INFLUENCE OF DIFFERENT KINDS
OF FOOD IN THE PRODUCTION OF BUTTER.

WE cannot, from a mere statement of the quantity of
the produce supplied to the dairy by a cow, judge of the
influence of any particular species of food upon the
animal, in consequence of the number of incidental cir
cumstances which tend to interfere with the natural pro
cesses carrying on in the animal system.   The present
series of experiments, as they have extended over a
longer period of time than any which have previously
been presented to the public, will tend in some measure
to exhibit irregularities dependent upon the conditions
in which the animals existed, and probably enable some
conclusions to be drawn explanatory of such apparent
anomalies.   It may be convenient to direct attention to
a few of these in considering some of the general conclusions.

I. *Quantity of Milk produced by different Kinds of
Food.*—In making inquiries respecting the amount of
milk afforded by cows, we cannot fail to be struck with
the vague and imperfect manner in which the attention

11*

of agriculturists is directed to weighing and measuring. Thus, for example, in Scotland, where milk is generally reckoned by the Scottish pint, when this measure is compared with the English system there is almost uniformly an error made in over-estimating its capacity. The usual allowance is four English to one Scottish pint ; but the true relation between these measures is much inferior to this—the English or imperial pint having a capacity of 34·659 cubic inches, and the Scottish pint of 103·4 cubic inches, a Scottish pint is very nearly equal to three English pints. When measurements have been made according to the Scottish system, a certain degree of caution must, therefore, be exercised in converting them to the English standard. Now, as in Scotland the actual measurements are generally made with the Scottish pints, when the amount of milk is stated in English pints we may almost safely conclude that the estimate has been greatly overdrawn ; but, even taking these sources of error into consideration, it is very remarkable how great a difference exists in the amount of milk given by cows under similar circumstances. No one will be surprised at the Alderney cow of Mrs. Tabitha Bramble* affording a daily supply of 4 gallons of milk, or 32 pints, when we read, in more recent times, of a short-horn giving 17 Scottish pints, (51 imperial pints,) or $64\frac{1}{2}$ lbs., at $10\frac{1}{4}$ lbs. to a gallon ; and of a roan cow yielding 30 Scottish pints, (90 imperial pints,) or $115\frac{1}{3}$ lbs., and requiring to be milked five times a day, so that at each milking $2\frac{1}{4}$

---

* " I am astonished that Dr. Lewis should take upon himself to give away Alderney without my privity and concurrants. Alderney gave four gallons a day ever since the calf was sent to market."—*Humphrey Clinker.*

gallons must have been extracted from the animal,* an average allowance for one cow during the whole day. All these statements must be understood as referring to cows which are allowed to graze at least during the day, and must be viewed as extraordinary cases. A nearer approach to an average will be obtained by directing attention to the produce of an Ayrshire cow fed in Berwickshire, which yielded, during July 1845, $6\frac{1}{2}$ Scottish pints, ($19\frac{1}{2}$ imperial pints,) 25 lbs. ; or to an Alderney cow in Lancashire, which supplied an average amount, in June 1845, of 20 imperial pints $= 25\frac{1}{2}$ lbs. ; but even in such instances, which are taken from low-land pasture grounds, the quantity often exceeds this by several pints, and sometimes also falls below it to the same extent, without any very apparent cause. In moorland pastures the average amount of milk is, how-ever, much inferior to what has been stated. In one locality in the neighborhood of Glasgow, where many cows are kept, the supply from each animal does not average more than from 12 ($15\frac{1}{2}$ lbs.) to 14 (18 lbs.) imperial pints per day ; and in another moorland farm the amount varies from 10 ($12\frac{3}{4}$ lbs.) to 15 (19 lbs.) im-perial pints. With a statement of these data for com parison we are enabled to form an idea of the influence exercised in the experiment detailed. When the cows were at pasture in Ayrshire they yielded 20 imperial pints each per day, ($25\frac{1}{2}$ lbs. ;) then they were in full exercise, and without any restriction in the amount of their food. They might in these circumstances be rep-resented as in a state of nature, and without any of the

---

* If the old Scotch wine measure is here meant, then it would be equivalent to about twelve imperial gallons.

Stephens' Book of the Farm, III. 1275.

artificial conditions which must always, to a certain
extent, interfere with the animal processes. An animal enjoying exercise must also consume a larger
amount of food than one shut up, or, in other words, it
must convey into the system a greater quantity of material for producing milk than an animal in a state of
confinement.

(1.) *Effect of Grass in producing Milk.*—For seven
days after coming to Glasgow, where they were confined in a roomy and airy cowhouse, and fed on cut
grass, the red cow (the less symmetrical of the two
animals) gave a larger amount of milk than when at
pasture ; the greatest quantity of milk during the week
being 27½ lbs., and the smallest amount being 24¾ lbs.,
the mean being 26⅓ lbs. ; there was therefore, in this
case, a decided increase in the amount of milk. With
the other cow the result was quite different ; the quantity of milk appears to have diminished immediately
with the confinement ; the mean of the first seven days
being 22¾ lbs. It is difficult to account for the great
difference in the result of the produce of the two animals upon any other supposition than that the constitution of the one admitted of confinement with less detri-
ment to its system than the other. The causes which
have been previously alluded to when treating of the
characters of the animals may, probably, also supply a
solution to these apparent anomalies. But we deduce
the important inference from these facts, that no correct
generalization can be arrived at from an isolated example. During the seven remaining days of the experiment the quantity of milk fell off with both cows
that of the brown cow subsiding from a mean of 26⅓

lbs. to 22½ lbs., and that of the white cow from 22¾ lbs.
to 20½ lbs.   There was, altogether, a difference in the
daily amount of milk, from the beginning to the end of
the fortnight, in the case of the brown cow of 4 lbs.,
and in the white cow of 2 lbs., although the amount of
food continued the same throughout.

(2.) *Influence of Variety of Food on Milk.*—The
considerable falling-off depended undoubtedly, in some
measure, upon the confinement to which the animals
were subjected, although on examining the tables it
will be found to be a pretty uniform result, that a change
of food produces an increase in the quantity of milk,
and that after the same diet has been continued for
some days the milk begins to diminish in amount.
There are several exceptions in the tables, some of
which, however, admit of simple explanation.   In the
second experiment, which was made with entire barley
steeped, the quantity of milk decreased very rapidly.
In the case of the brown cow there was a difference
between the milk of the first and last day of the experiment of 5 lbs., and in the white cow of 2½ lbs.   This
arose from a quantity of the barley being ejected by the
animals without being digested.   Entire malt being
given raised the amount of milk immediately, and the
quantity continued to rise daily till it amounted at the
end of the trial, in the case of the brown cow, to an
increment of the last over the first day's milk of 3 lbs.,
and in the white cow of 4 lbs.   We can see at once
why there was an improvement under the malt regimen, from the circumstance that, being much more
soluble than the barley, it was not ejected by the animals ; indeed, none of it was observable in the dung,

while a considerable proportion of barley was always carried to the dung-heap. The second and third experiments do not serve to prove any point in reference to the dietary of animals, but they may be useful as evidence to show that the more divided the food is, the greater is the amount of milk produced. In the fourth experiment, with crushed barley, the brown cow's milk decreased 1¼ lbs. in sixteen days, and the white cow's 10 oz., or considerably more than half a pound, in the same period. In the fifth experiment, with crushed malt, the brown cow's milk declined 2¼ lbs. in sixteen days, and the white cow's upwards of 2¼ lbs. In the sixth experiment, with a larger quantity of crushed barley, the brown cow's milk continued to increase up to the fourth day, and then began to decline ; a similar result attended that of the white cow. In the seventh experiment, with molasses and barley, the brown cow's milk reached its acme or culminating point on the second day of the trial, and it then continued to decline till the close of the experiment on the tenth day. With the white cow, the greatest amount of milk was afforded on the fifth day, when it began to decline and gradually diminish till the termination of the trial. In the eighth experiment, made with barley and linseed, the amount of milk continued to increase for a longer period than usual ; the largest quantity given by the brown cow was on the ninth day, and by the white on the eighth and ninth days. With the bean meal, in the ninth experiment, the milk continued to increase up to the fifth day, when the trial closed.* That a change of diet is necessary for animals which are kept in a confined

---

* See Diagram, and Miscellaneous Table No. IV.

condition is proved by the tables previously given, in a striking manner, and the results now obtained amply sustain the idea supported by me some time ago in reference to the dietary of human beings shut up in poor-houses and places of confinement. It was then argued that, " in order to retain the human constitution in a healthy condition, variety of food should be properly attended to,"[*] and different species of diet were suggested as well calculated to supply a series of dishes to the poor. In the Asylum for the Houseless, and in the House of Refuge at Glasgow, the recommendations were followed out ; and, according to the report of the treasurer, Mr. Liddell, the dinner meals being varied two or three times every week, "the change in the dietary routine is much relished by the inmates, and may have had some effect in the greater degree of health which has been evident among them of late."[†]

---

[*] Proceedings of the Philosophical Society of Glasgow, p. 39.

[†] Proceedings of the Philosophical Society of Glasgow, vol. i. p. 40.

The following economical and wholesome dishes are formed on the principles enunciated, and are used in the public charities of Glasgow

Fish Pudding for Ten Persons.

Quantity for One.		Quantity for Ten.			s.	d.
2 lbs.	0 oz.	20 lbs.	0 oz.	potatoes, at $\frac{1}{4}d$. per lb. - -	0	5
0	8	5	0	salt fish, at 2d. per lb. -	0	10
0	0$\frac{1}{4}$	0	2$\frac{1}{2}$	lard or dripping, at 8d. per lb. -	0	1$\frac{1}{2}$
				pepper - - - -	0	0$\frac{1}{2}$
2	8$\frac{1}{4}$	25	2$\frac{1}{2}$		1	5

Cost, exclusive of fire and cooking, under 1$\frac{3}{4}d$. per head. Steep and boil the fish as long as the saltness and size of the article to be used requires, take out the bones, boil the potatoes in a separate vessel, beat the whole together. If a fire or oven can be had, brown the top of the dish.

The analogy subsisting between the physical nature of human beings and of many of our domestic animals would lead us to the conclusion, upon physiological grounds, that their dietary should be conducted upon precisely similar principles. To prove this by exact experiments is a point, it will be admitted, of considerable importance to the agriculturist, although it may have been, as might be expected, surmised by many intelligent observers. Not only, however, is *variety* of food requisite for an animal in an artificial state, it is found also to be beneficial to one in a condition more akin to that of nature. For it is upon this principle

A Stewed Hash of Sheep's Draught for Ten Persons.

Quantity for One.		Quantity for Ten.			s.	d
2 lbs.	0 oz.	20 lbs.	0 oz.	potatoes, at ¼d. per lb.	0	5
0	5½	3	8	two sheep's draughts, 5d each	0	10
0	0	0	8	onions, 1d.; pepper, salt, and flour, 2d.	0	3
2	5½	24	0		1	6

Cost, exclusive of fire and cooking, full 1¾d. per head. Boil the lights for one hour, preserving the water; hash said lights, liver, and heart together with flour, pepper, salt, and onions; then stew the whole for one hour, using the water in which the lights were boiled. The boiling and stewing should be done over a very slow fire.

A Mince of Cow's Heart for Ten Persons

Quantity for One.		Quantity for Ten.			s.	d.
2 lbs.	0 oz.	20 lbs.	0 oz.	potatoes, at ¼d. per lb.	0	5
0	4	2	8	half a heart, 1s. 6d. -	0	9
0	0	0	8	onions, 1d.; pepper, salt, and flour, 1d.	0	2
					1	4

Cost, exclusive of fire and cooking, full 1½d. per head. Cut up and wash the heart well. Mince it very small, using onions, flour, pepper, and salt. Stew the whole over a slow fire for two hours.

that we are able to account for the superior influence
of old natural pastures, which consist of a variety of
grasses and other plants, over those pastures which are
formed of only one grass, in the production of fat cattle
and good milk cows. To any one who considers with
attention the experiments which have been detailed,
there cannot remain a doubt in the mind that cattle,
and especially milk cows, in a state of confinement
would be benofited by a very frequent and entire
change in their food. It might not be too much to say
that a daily modification in the dietary of such animals
would be a sound scientific prescription. The effect
of variety of food is exhibited in the frontispiece. In
considering the case of the white cow, we find that a
change from barley to barley and molasses increased
the milk in three days from 21 lbs. 6 oz. to 23 lbs. 7 oz.;
on changing from malt to barley it increased from 19
lbs. 10 oz. to 20 lbs. 11 oz. on the first day; from barley to barley and linseed, it increased from 21 lbs. 2 oz.
to 23 lbs. 12 oz. on the sixth day ; from barley and linseed to beans, it increased on the first day from 21 lbs.
13 oz. to 23 lbs. 14 oz. Some of these changes can
be traced in the diagram placed as a frontispiece, while,
at the same time, we obtain from it a distinct view of
the relative influence of the different species of food in
keeping up a great or regular supply of milk.

(3.) *Effect of Barley and Malt on Milk.*—In considering the influence of barley and malt on the production of milk, it is obvious that Experiments II. and
III. offer no data from which conclusions can be drawn,
except to point out the useful practical fact, that grain
should never be given to cows in an entire state, but

12

that it should always be ground or crushed, and then steeped before being presented to them. If we compare experiments IV. and V., we find that in sixteen days 141 lbs. of crushed barley steeped produced in the brown cow 342 lbs. of milk, and in the white 351 lbs. of milk, and that both animals gained in weight; while, again, 168 lbs. of malt produced in the brown cow 310 lbs. of milk, and in the white 345 lbs. of milk, during sixteen days; the former cow gaining some weight, and the latter losing a little. The quantity of malt exceeded that of the barley by 27 lbs., and yet the brown cow gave 32 lbs. less of milk with malt than with barley, and the white cow only 6 lbs. less milk; hence, in the brown cow 100 lbs. of barley produced as much effect as 131 lbs. of malt, and in the white cow 100 lbs. of barley were equivalent to 119 lbs. of malt. Now, as 100 parts of barley, when malted, become eighty of malt, it is obvious that 100 parts of barley are equal in value to 125 of malt, for 80 : 100 :: 100 : 125. If we take the mean of the result of the preceding experiment, we find that 100 of barley go as far in producing milk as 125 of malt, $119 + 131 \div 2 = 125$. Again, by a mean of three experiments, the amount of nitrogen in malt was found to be 1·52 per cent., and that of barley 1·96 per cent., by four experiments, which would make 100 parts of barley equivalent to 128 of malt in nutritive power. These are all remarkable coincidences of theory and practice, and cannot fail to convince us that the proportions stated are very close approximations to the nutritive equivalents of barley and malt, or, in other words, that malt is about one-fifth inferior to barley in its nutritive effects. In considering the sixth experiment, which was made for the purpose of comparing the effect of a large quantity of barley with a large amount of malt, it will be observed, that the experiment commenced when the amount of milk was declining under the malt regimen, but that as soon as the barley was given the milk began to increase in both cows. The weather, however, at this time, became much warmer than it had hitherto been. The mean temperature, as exhibited in the table, became more elevated, but the numbers in the table will scarcely give an idea of the stagnant sultry nature of the atmosphere in the cowhouse, in the immediate neighborhood of which, in a room without a fire, the thermometer during the five days stood at 66°, and at one period of the thirtieth, or first day of the experiment, rose to 70°. The cattle were, during this period, very much troubled with flies, which produced, as all agriculturists will understand, much agitation and constant movement. These circumstances are calculated to explain the loss of weight sustained by the brown cow, and they account for the fact that the increase of milk was not so rapid as in the previous barley experiment. This experiment may be viewed as an interesting example of the influence which atmospherical causes exercise upon the production of milk, and exhibits a result perfectly in accordance with the experience of good agricultural observers. From the circumstances mentioned it is obvious that this experiment should not be taken apart from the previous barley trial, since the conditions were somewhat different under which it was made; but we have employed it along with the other trial in striking an average, as in Miscellaneous Table No. IV. Another effect which came into operation in this experiment I believe to be, that the quantity of barley was too great, and that more nutritive matter was given in proportion to the heat-producing matter than was fitted for the support of the system, and thus gave occasion to a deteriorating action.

(4.) *Effect of Molasses, Linseed, and Beans in the Production of Milk.*—If we examine the Miscellaneous Table No. IV., we find the mean quantity of milk afforded by the brown cow, every five days under different regimens, was as follows :—Barley, 107 lbs. ; malt, 97 ; barley and molasses, 101 ; barley and linseed, $102\frac{1}{2}$; beans, $99\frac{3}{4}$. And by the white cow the mean quantities respectively were, every five days, barley, 109 lbs. ; malt, $108\frac{1}{3}$; barley and molasses, $112\frac{1}{4}$; barley and linseed, $115\frac{1}{2}$; beans, $115\frac{6}{10}$. Of all these articles of food, in both cases, malt gives the smallest produce. Then comes, with the white cow, barley, and the other articles increase in effect as they stand above, bean meal affording the greatest amount of produce. It will be observed, in examining the bean meal table, that the milk increased up to the termination of the experiment ; and that in the case of the white cow, the quantity yielded exceeded that supplied by this animal on any previous occasion, except in one solitary instance under the grass diet. The quantity of milk given by the white cow on the 18th September, under the bean regimen, amounted nearly to $25\frac{1}{3}$ lbs., thus approaching closely to that afforded by both cows when they were at pasture three months previously. This cannot fail to be admitted as an interesting fact, and is strongly corroborative of the propriety of the partiality of cow-feeders for bean meal as an article of nutrition for their stall-fed cattle. If we take a mean of the produce of the two cows, as previously stated, we shall find the relative influence of each in the production of milk to be as follows :—Commencing with that which possesses the lowest nutritive power, malt produces 102·66 lbs. of milk; barley and molasses, 106¾; bean meal, 107·68; barley, 108; barley and linseed, 109. We think it better to state the mean produce of the two cows, because it will afford an average of what we might expect to meet with in feeding a number of cattle with these various articles of food. A comparison of the experiments on the two cows, however, fully demonstrates that one kind of food will produce a greater influence on one animal than on another; and that, as with human beings, probably, attention should be bestowed on what is agreeable to each individual animal, both in reference to its palate and constitution. For it should be always borne in mind that stall-fed animals are not in a natural condition, and that being placed under artificial restrictions, a due consideration of the adequate means of counterbalancing the adverse circumstances of their condition can alone conduce to a true theory of humane stall-feeding.

(5.) *Influence of Quantity of Grain in the Production of Milk.*—To ascertain the amount of grain best calculated to afford the largest supply of milk is a practical point of no small importance to the cow-feeder. Perhaps from Miscellaneous Table No. IV. the best solution to this question may be obtained, in reference to the articles of food employed in the present series of experiments. In the barley experiment it will be observed, that when 12 lbs. of barley were given daily, the amount of milk was inferior, in both cows, to that

12*

obtained when 9 lbs. was the diurnal allowance. This result seems so decided, in both series of experiments, that it may almost be considered as established, that no adequate advantage appears to be attained by pushing the supply of barley to a cow beyond the extent of 9 lbs. daily. An increase in the quantity of malt appears sometimes to increase the quantity of milk ; but, in general, the same deduction may be made with reference to malt as to barley, that in a remunerative point of view, 9 lbs. a day may be considered a larger proportion of malt to supply a cow. It is highly probable, indeed, that a smaller amount, especially if the animals were allowed a certain limited degree of exercise, would be found fully as efficient as a larger quantity. We have, in the body of the report, endeavored to explain this upon the physiological principles of digestion, and to show, that, as ruminating animals more especially are possessed of great capacity of stomach, an excess of concentrated food, by failing to effect adequately the purpose which bulky food accomplishes— of exciting the coats of the stomach to secrete their digesting fluid—will tend rather to diminish than to increase the result which we desire to gain.

(6.) *Rate at which Food is changed into Milk.*—As a variety of views prevail with regard to the period required by the animal system for the conversion of food into milk, I endeavored to solve this question by keeping an accurate register of the amount of milk supplied by a cow, morning and evening. From this register it appears, that in the course of a month, the brown cow gave the largest amount of milk in the evening only 6 times, while the white cow was in the same condition only 3 times. It may be considered therefore certain, during these experiments, that as a general rule the greatest quantity of milk was yielded by the cows in the morning. An example, taken at random from the register of the white cow's milk, will show the force of this observation :—

Food.		Milk.		
		lbs.	oz.	drs.
Aug. 1. Barley and hay	morning	11	8	15
	evening	10	3	14
2. —	morning	11	7	1
	evening	9	11	9
3. —	morning	11	10	15
	evening	9	11	9
4. —	morning	10	14	5
	evening	9	11	3
5. Barley, molasses, and hay,	morning	11	4	10
	evening	10	4	8
6. —	morning	12	5	7
	evening	10	10	11

Now, as comparatively a small amount of food is consumed during the night, it is obvious that this superior amount of milk must be derived from the previous day's fodder. An observation which was frequently made, viz. that undigested food did not appear in the dung till sixteen hours after being swallowed, would tend to demonstrate that, during this period at least, absorption of the nutritive part of the food was going on ; since we know that along the whole course of the intestinal canal the soluble food continues to be taken up through the coats of the viscera.

II. *Relative Influence of different Kinds of Food in the Production of Butter.*—In the Table IV. (Appendix) we have collected the amount of butter produced by five kinds of food during periods of five days each. But previous to these trials, thus arranged, the largest quantity given by the brown cow was under the grass regimen. The first five days of the experiment yielded 4·93 lbs. of butter, after which the quantity diminished to the last five days of the trial, when the quantity yielded amounted to 3·75 lbs., a proportion not superior to what was produced in some of the subsequent experiments. The same law does not appear to hold with reference to the diminution of the butter as pertains to that of the milk, when the food has been continued for some time. We find, on the contrary, frequently the amount increasing towards the close of the experiment, even when it is continued for ten or fifteen days. The largest amount of butter was afforded in the brown cow by crushed barley. During the third series of five days the amount was 3·935 lbs. ; bean meal gave the next greatest quantity 3·69 lbs. in five days ; then comes barley and linseed, 3·689 lbs. during the first five days ; barley and molasses, 3·63 lbs., and malt 3·60 lbs. In the case of the white cow the quantity was, beans, 3·76 ; barley and linseed, 3·421 ; crushed barley, 3·376 lbs. ; barley and molasses, 3·26 ; and malt 3·126. With both animals we observe that malt is lowest in the scale, a fact which seems in some measure to militate against the idea of the origin of the butter being in the sugar of the food. Be this as it may, however, although there are many counter arguments in favor of the opinion that sugar affords such a supply, we think the Tables II. and III. (Appendix) tend to show that there is no relation between the butter of the milk and the wax and oil of the food ; since frequently, when the oleaginous matter of the food is small, the butter is more considerable than on other occasions when the reverse happens. Since then the facts contained in the tables, and the arguments used in the body of the report, seemed to prove that the butter cannot be supplied from the oil of the food, it becomes an interesting point for the agriculturist to learn from what element of the food it proceeds. It may safely be inferred that it must be formed from some other constituent of the diet by means of the vascular system, either as a primary or secondary stage. Sugar affords the most simple element from which it may be produced, because we now understand how the acid of butter can originate from sugar ; but even the albuminous principles might afford butter. (*Würtz*, Liebig.) Upon these grounds, then, we can infer that a certain degree of exercise would be more conducive to the production of fat than if the animal is allowed to remain at rest ; because, as the source of the fat or butter is dependent on the process of respiration, it is obvious that the more the function is encouraged within moderate bounds, the greater will be the amount of the oil-giving principle of the food taken into the system and converted into fat. We believe that this theoretical deduction is perfectly in consonance with the experience of good observers, who find that box or hammel feeding is more conducive to health of cattle and cows destined for the butcher, or for the production of butter, than close plant-like confinement, which is foreign to the nature of every animal, and at variance with the first principles of physiological science.

It appears to result from these experiments, as an irresistible conclusion, that the fat or butter of the milk

must be produced at the expense of the calorifient ingredients of the food, aided by the presence of the nutritive or azotized principles ; and that the greatest product of butter must be obtained when the two ingredients of the food are present in the best proportions.

CHAPTER IX.

MUSCLE OF THE BODY SUPPLIED BY THE FIBRIN OF THE FOOD.—FIBRIN
SUPPLIES HEAT TO THE BODY.—ADDITIONAL OR CALORIFIENT FOOD AL-
SO REQUIRED.—AMOUNT OF NUTRITIVE AND CALORIFIENT FOOD CON-
SUMED BY A COW PER DAY.—THE TRUE LAWS OF DIETING.—AMOUNT
OF NUTRITIVE MATTER IN VARIOUS KINDS OF VEGETABLE FOOD.—
ARROW-ROOT IMPROPER FOR INFANT FOOD, BUT USEFUL IN DISEASES.
—THE LARGEST QUANTITY OF MILK PRODUCED BY FOOD CONTAINING
THE GREATEST AMOUNT OF NITROGEN.—GRASS AN EXCEPTION TO THIS
RULE.—EXPLANATION OF THIS FACT.—NEW FORMS OF BREAD.—OAT-
MEAL BREAD—BARLEY BREAD—INDIAN CORN BREAD—PEAS BREAD.—
MODE OF BAKING.—DIFFERENCE BETWEEN FERMENTED AND UNFER-
MENTED BREAD.—UNFERMENTED BREAD RECOMMENDED.

THE idea which is now entertained by physiologists,
that the muscular part of the animal frame is derived
from the albuminous constituent of the food, was clearly pointed out by Beccaria in the year 1742. (*Histoire
de l'Académie de Bologne*, Collect. Acad. xiv. 1.) He
demonstrated that the flour of wheat contained two
characteristic ingredients, which on distillation or digestion afford products totally dissimilar to each other.
One of these, which he termed the starchy part, resembles in its constitution vegetables, and supplies
analogous products. Vegetable substances, he says,
may be recognised by their fermenting, and yielding
acids without exhibiting symptoms of putrefaction. The
glutinous part of flour, on the contrary, resembles animal matter, the distinguishing feature of which is its
tendency to putrefaction and conversion into a urinous
(ammoniacal) liquid. " So strong," he adds, " is the

resemblance of gluten to animal matter, that if we were not aware of its being extracted from wheat, we should not fail to mistake it for a product of the animal world." To be convinced that he considered this identical substance to enter into the constitution of our frames, it is only necessary to quote his query: "Is it not true that we are composed of the same substances which serve as our nourishment?" The same doctrine has been taught and practised in our own country, in more recent times, by Dr. Prout, and is now almost universally received by European physiologists, although the true authors may not have been always recognised. That the systems of animals are capable of sustentation by a supply of fibrinous matter almost alone is obvious from the history of the primitive inhabitants of the prairies of America. It is stated on good authority, (Catlin,) that there are 250,000 Indians who live almost exclusively on buffalo flesh during the year. The fresh meat is cut in slices of half an inch in thickness across the grain, so as to have fat and lean in layers, and is hung up exposed to the sun and dried. Upon this food, which is pounded, and eaten sometimes with marrow, the wild hordes of the West are not only nourished, but it is obvious that the heat of their bodies is kept up, since they taste no vegetable food whatever. Fibrin, then, is calorifient, or capable alone, we infer, of producing animal heat. Liebig, it is well known, divides the functions of the food into nutritive and respiratory. I have ventured to employ, instead of the latter term, the expression *calorifient* or *heat-producing*, so as to give a wider range through the whole system to the function of the unazotized food than the more local term of respiratory would appear to imply. According to this view all food is destined for repairing the waste of the body, and for the production of animal heat. The heat may be produced by the union of the carbon and hydrogen of the food with oxygen (the latter gaining admission to the system by the lungs, stomach, and skin,) or by the condensation of oxygen during its substitution for hydrogen and formation of oxygen products. The preceding inference we also deduce from the experiment in which a dog was fed for some weeks on the glutinous matter of flour, (*Magendie ;*) and it may be further concluded, that fibrinous or albuminous matter when given alone is partially converted into carbonic acid, and is removed from the system during the process of expiration. But it would appear, from consideration of the experiments which have been made on the nutrition of animals with pure fibrin, that an auxiliary in the production of animal heat is either indispensable or highly advantageous, since animals fed on fibrin alone invariably declined in health, (Magendie,) and the American Indians have a certain admixture of fat with their dry meat, and are in the habit likewise of using marrow with it.

The reason why an auxiliary is required for the supply of animal heat appears to be, that the fibrinous matter which is taken up by the vessels of the intestines and is carried into the blood, requires to pass through the condition of muscular tissue before it can be of service as a calorifient agent. The only view which appears at present to be tenable is, that all or the greater part of the fibrinous and albuminous matters which enter the blood displace a certain amount of the same substances existing in a solid form in the system, as brain, muscle, &c., and that the displaced matter undergoes certain modifications; probably, for example,

13

it passes into the form of gelatin, and is excreted in the soluble state of urea, uric acid, and nitrogenous products. It is in passing into these last conditions that we can alone expect fibrinous matter to give out animal heat. Time, therefore, is required to produce these changes. It is to save the system, then, from too rapid waste, and at the same time to afford an abundant supply of heat, that the calorifient food is required, and is always employed by all members of the human family who have advanced beyond the savage state.

That the amount of calorifient food, in contradistinction to nutritive food, properly so called, as it has been well defined by Liebig, is out of all proportion greater than that required to supply the waste of the solid matter of the body, is obvious from the following table, which represents the amount of the ultimate constituents of the food of a stall-fed cow, consumed during one day :—

	Food.	Fæces.	Consumption.
	lbs.	lbs.	lbs.
Carbon -	11·90	5·10	6·80
Hydrogen	1·61	0·62	0·99
Nitrogen	0·15	0·20	0·25
Oxygen	10·74	4·12	6·62
Ash	1·71	1·09	0·62
	26·41	11·13	15·28

The food in this case was grass, (the *lolium perenne*, or rye-grass.) If we now calculate the amount of food which was destined for nutrition by the formulæ below,*

---

* Albuminous matters contain about 53 per cent. of carbon, 7 of hydrogen, 16 nitrogen, and 24 oxygen. Hence, to obtain the carbon

we find that it amounts only to 1·56 lbs., as represented
in a tabular form :—

	Nutritive. lbs.		Calorifient. lbs.
Carbon - -	0·828		5·982
Hydrogen -	0·109	- -	0·771
Nitrogen	0·250	-	
Oxygen	0·373	-	6·247
	1·560		13·000

A true system of dieting would therefore require such
tables for each condition of animals, in order that a
comparison may be instituted between the wants of the
system and the food. If this mode of viewing the
question be correct, then the relation of the nutritive
part of the food absorbed by the animal system in the
preceding experiment is to the calorifient portion as 1
to $8\frac{1}{3}$ nearly. By comparing this fact, then, (which is
independent of all hypothesis,) with the different varieties of human food, it is probable that some light may
be obtained in reference to the differences in the relative proportion of these constituents. Milk, for example, the food of the infant mammalia, contains one part
of nutritive to two parts of calorifient constituents, and
in the growing state of an animal the nutritive part of
the food not only supplies the place of the metamor-
phosed solids; but an additional amount of it is required
to increase the bulk of the individual ; and, as we have
already stated that animal heat is generated by the
change or degradation of the fibrinous tissues, it is

---

from the above table we have $\dfrac{·25 \times 53}{16}$ = ·828 lbs. carbon; for the
hydrogen $\dfrac{·828 \times 7}{53}$ = ·109 lbs. hydrogen ; for the oxygen $\dfrac{·109 \times 24}{7}$
= 0·373 lbs. oxygen.

obvious that in the nourishment of infant life there is a
supply of heat from the casein, vastly superior to that
afforded by fibrin supplied to full-grown animals, because the amount taken in proportion to the quantity
of calorifient matter is much greater. If we refer,
again, to the food which is generally employed by the
inhabitants of this country, wheat and barley, we find
by a mean of experiments afterwards to be detailed,
that the average amount of albuminous matter present
in them is 11 per cent., while the quantity of starch and
sugar existing in these substances may vary from 70 to
80 per cent.; thus affording the proportion of nutritive
to calorifient food as 1 to 7, and upwards. Such food,
it may be inferred, is fitted for the consumption of an
animal which is not subjected to much exercise of the
muscular system, and may be viewed as the limit of
excess of the calorifient over the nutritive constituents
of food. As the demands upon the muscular part of
the frame become more urgent, the proportion of the
azotized or nutritive constituents should be increased,
and this may be extended until we arrive at the point
where the fibrinous matter is equal to the half of the
calorifient, which is probably, in a perfectly normal
physiology, the greatest relative proportion of nutritive
material admissible.

The proportion of the nutritive to the calorifient
constituents of food should therefore vary according as
the animal is in a state of exercise or rest; and it is
upon the proper consideration of such relations that the
true laws of dieting depend. For calculations of this
nature, tables exhibiting the amount of albuminous
matters in the different articles of food are indispensable, as they afford at a glance the required knowledge.

The constituents of the flours used as human food are principally albuminous matter, calorifient matter, water, and salts ; so that when we have determined the amount of albuminous substance in the dried condition of the flour, the remainder may be estimated as calorifient matter without any sensible error. In the following table the water has not been removed from the flour The numbers are the results of my experiments :—

	Albuminous or Nutritive Matter per cent.
Bean meal	25·36
Linseed meal	23·62
Scotch oatmeal	15·61
Semolina	12·81
Canadian flour	11·62
Barley	11·31
Maize	10·93
Essex flour	10·55 to 11·80
East Lothian flour	9·74 to 11·55
Hay	9·71
Malt	8·71
Rice (East Indian)	8·37
Sago	3·33
South Sea arrow-root	3·21
Tapioca	3·13
Potatoes	2·23
Starch (wheat)	2·18
Swedish turnips	1·32

The numbers represent the amount of albuminous matter contained in 100 parts of the various substances as they occur in commerce. As all of the substances in the table contain from 5 to 14 per cent. of water, certain deductions are required, to arrive at the true amount of calorifient matter. In general, it may be stated that wheat flour, maize, barley, and beans contain from 10 to 14 per cent. of water, while oatmeal contains 6 per cent., and tapioca, arrow-root, and sago from 10 to 13 per cent. In order to arrive at the true amount of calorifient matter contained in the substances in the table, we have only to deduct the amount of albuminous substances, with the water and salts, which, upon an average, amount together to about 12 to 15 per cent. Then, by dividing the remainder, or calorifient matter, by the amount of albuminous substances, we obtain the relation subsisting between the nutritive and calorifient constituents. In this manner tables may be constructed, illustrating the true practice of dieting.

Approximate Relation of Nutritive to Calorifient Matter.

	Relation of Nutritive to Calorifient.
Milk.—Food for a growing animal	1 to 2
Beans	1 — 2½
Oatmeal	1 — 5
Semolina } Barley }	1 — 7
English Wheat Flour.—Food for an animal at rest	1 — 8
Potatoes	1 — 9
Rice	1 — 10
Turnips	1 — 11
Arrow-root } Tapioca } Sago }	1 — 26
Starch	1 — 40

From this table we are led to infer that the food destined for the animal in a state of exercise should range between milk and wheat flour, varying in its degree of dilution with calorifient matter according to the nature and extent of the demands upon the system. The animal system is thus viewed as in an analogous condition to a field from which different crops extract different amounts of matter from the soil, which must be ascertained by experiment. An animal at rest consumes more calorifient food in relation to the nutritive constituents than an animal in full exercise. The food, therefore, employed by a person of sedentary habits should contain more calorifient and less nutritive matter than one whose occupations cause him to take more exercise. It is to be desired that some light should be thrown on this subject by careful experiments. The food of animals and the manure of plants we thus see afford somewhat of a parallelism. Milk may therefore be used with a certain amount of farinaceous matter, such as the class of flours and meals, with probable advantage; but the dilution should not exceed the prescribed limits. It is thus that we may explain the fact of beans, oats, oatmeal, and barley meal being used so extensively in the feeding of horses. These articles of food, however, do not suffice alone : calorifient matter in the form of hay should also be administered. From this table, likewise, we infer that, as nature has provided milk for the support of the infant mammalia, the constitution of their food should always be formed after this type. Hence we learn that milk, in some form or other, is the true food of children, and that the use of arrow-root, or any of the members of the starch class, where the relation of the nutritive to calorifient matter is as 1 to 26 instead of being as 1 to 2, by an animal placed in the circumstances of a human infant, is opposed to the principles unfolded by the preceding table. In making this statement, I find that there are certain misapprehensions into which medical men are apt to be led at the first view of the subject. To render it clearer, let us recall to mind what the arrow-root class of diet consists of. Arrow-root and tapioca are prepared by washing the roots of certain plants until all the matter soluble in water is removed. Now, as albumen is soluble in water, this form of nutritive matter must in a great measure be washed away: under this aspect we might view the original root before it was subjected to the washing process, to approximate in composition to that of flour. If the latter substance were washed by repeated additions of water the nitrogenous or nutritive ingredients would be separated from the starchy or calorifient elements, being partly soluble in water, and partly mechanically removed. Arrow-root, therefore, may be considered as flour deprived as much as possible of its nutritive matter. When we administer arrow-root to a child it is equivalent to washing all the nutritive matter out of bread, flour, or oatmeal, and supplying it with the starch; or it is the same thing approximately as if we gave it starch; and this is in fact what is done, when children are fed upon what is sold in the shops under the title of farinaceous food, empirical preparations of which no one can understand the composition without analysis. Of the bad effects produced in children by the use of these most exceptionable mixtures, I have had ample opportunities of forming an opinion, and I am inclined to infer that many of the irregularities of the bowels, the production of wind, &c., in children, are often attributable to the use of such unnatural species of food. How often are the ears of parents and nurses distressed with the agonizing cries of the helpless child, and how often are these symptoms of suffering treated as the effects of ill-humor, or of causeless peevishness; when, on the contrary, they have been produced by the improper diet in many cases with which the child has been supplied ! It should be remembered that all starchy food deprived of nutritive matter is of artificial production, and scarcely, if ever, exists in nature in an isolated form. The administration of the arrow-root class is therefore only admissible when a sufficient amount of nutritive matter has been previously introduced into the digestive organs, or when it is inadvisable to supply nutrition to the system, as in cases of inflammatory action. In such instances the animal heat must be kept up, and for this purpose calorifient food alone is necessary. This treatment is equivalent to removing blood from the system, since the waste of the fibrinous tissues goes on, while an adequate reparation is not sustained by the introduction of nutritive food. A certain amount of muscular sustentation is still, however, effected by the use of arrow-root diet; since, according to the preceding tables, it contains about one-third as much nutritive matter as some of the wheat flours. The extensive use of oatmeal, which is attended with such wholesome consequences among the children of all ranks in Scotland, is, however, an important fact deserving of serious consideration ; and, it appears to me, is strongly corroborative of the principles which I have endeavored to lay down in the preceding pages. After the explanations which have been given, it is scarcely necessary to particularize further the specific nature of the food to be recommended for the use of children. A certain admixture of milk, the natural type of the food, is still to be retained, while the solid matter to be prepared along with it may be of great variety, such as bread made into panado, semolina or pounded wheat ;

I believe this kind of food, which is sold in the shops, to be generally prepared from wheat brought from a more temperate region than that of this country, in consequence of the amount of nitrogen which I have found in it. The best American wheat flour, good Scottish oatmeal, and barley-meal, may all be employed at different times by way of variety, and repeated according to their agreement with the child's organs of digestion. The digestion of all these forms of food containing starch is greatly promoted by long boiling either with water or milk, as this process is just so much labor saved to the intestinal organs. It is thus obvious that we have a great variety of food fitted for children of which we know the composition, and that we should prefer it to any species of compounded stuff the constitution of which we are ignorant. It is a sufficiently remarkable fact, that oats increase in nutritive power in proportion to the increase of latitude within certain limits, while wheat follows an inverse law. Those who are in the habit of representing mankind as the " lords of the creation," who take the limited view of considering all that we see around us as created merely for their use, misapplying the thought—" the proper study of mankind is man ;" and who thus, with the characteristic vanity of earthliness, follow the footsteps of Kant, profanely attempting to survey the divine mind, will discern probably in this curious circumstance further proofs of their theory, as if to show " how little can be known."

In the table which contains the amount of albuminous matter in different kinds of food, a second column, in accordance with tables of this description, might have been added, representing 100 parts of beans as equal in nutritive power to 1160 of starch; but if the views now explained are legitimate, we see that such a method of estimating nutritive power is not founded on scientific principles. In a correct plan of dieting the proper equilibrium must be retained between the demands of the animal organism and the constitution of the food, otherwise, either the nutritive or calorifient system must be deteriorated. These views sufficiently explain the experiments which have been made upon cows; in which the result was unfavorable, when they were fed on potatoes and beet-root in considerable quantities, as both of these substances contain an excess of calorifient matter. It is well known to feeders of cattle, that an animal fed on large quantities of potatoes is liable to complaints, such as affections of the skin, and also to loss of weight. These consequences, it may be readily inferred, are derived from the want of the proper balance between the elements of the food.

The importance of attention to the proper equilibrium of the constituents of the food is clearly pointed out in the following table, from which it is evident, that food containing the greatest amount of starch or sugar does not produce the largest quantity of butter, although these substances are supposed to supply the butter; but the best product of milk and butter is yielded by those species of food which seem to restore the equilibrium of the animals most efficiently. The first column in the table represents the food used by two cows; the second column gives the mean milk of the two animals for five days; the third, the butter during periods of five days; while the fourth contains the amount of nitrogen in the food taken by both animals during the same periods:—

		Milk in five Days.	Butter in five Days.	Nitrogen in Food in five Days.
		lbs.	lbs.	lbs.
I.	Grass	114	3·50	2·32
II.	Barley and hay	107	3·43	3·89
III.	Malt and hay -	102	3·20	3·34
IV.	Barley, molasses, and hay	106	3·41	3·82
V.	Barley, linseed, and hay -	108	3·48	4·14
VI.	Beans and hay - -	108	3·72	5·27

We may infer, from these results, that grass affords the best products, because the nutritive and calorifient constituents are combined in this form of food, in the most advantageous relations. The other kinds of food have been subjected to certain artificial conditions, by which their equilibrium may have been disturbed. In the process of hay-making, for example, the coloring matter of the grass is either removed or altered; a portion of the sugar is washed out or destroyed by fermentation, while certain of the soluble salts are removed by every shower of rain which falls during the curing of the hay. Perhaps similar observations are more or less applicable to the other species of food enumerated.

The principles which we have been endeavoring to explain being understood, little difficulty will be experienced in constructing dietaries, so as to meet the wants of the animal system under the particular circumstances in which it may be placed. By various mixtures of one kind of flour, less supplied with azotized matter, with another which is richer in this material, the equilibrium of the food which from meteorological causes prevailing in any particular country, may not have reached the proper standard, may be effectually restored. The wheat of England, for example, is inferior to that of the continent of Europe, and of America, as appears from the table. It may, however, be- improved by an admixture either with foreign flour, or with oatmeal, barley, beans, or any of those substances which stand above it in the table; and in this state it will be found to form palatable bread. All these species of grain owe their nutritive properties to the presence of fibrin, casein, gluten, and albumen. It is in the predominance of gluten over the other azotized materials that wheat owes its superior power of detaining the carbonic acid engendered by fermentation, and thus communicating to it the vesicular spongy structure so characteristic of good bread. By mixing one-third of Canada flour with two-thirds of maize, a very good loaf is produced, and when equal parts of flour and oatmeal, or of barley, or of peasmeal, are employed, palatable bread is the result. Beneficial effects would probably follow from the admixture of two or three different kinds of grain, and many of these forms of bread might be substituted with advantage for pure wheat flour in peculiar conditions of the system.

When it is proposed to make a loaf of oatmeal and flour, the common oatmeal should be sifted so as to obtain the finest portion of the meal, or it may be ground to the proper consistence. This should be mixed then with an equal weight of best flour, Canadian, for example, and fermented. I have not succeeded in making a good loaf with a smaller amount of flour than one half, although I have tried it in various proportions. If we were to attempt to raise oatmeal without an admixture with flour, in consequence of the absence of gluten that principle which retains the carbonic acid of fermentation, we should obtain only

14

a sad, heavy, doughy piece of moist flour. This form of bread, it appears to me, and to many who have ex amined it, would be a great improvement on the hard, dry oat-cakes, so much used in the more unfrequented parts of our country, where the inhabitants have scarcely as yet commenced to share in what are in other localities considered to be necessaries of life. It is an observation which all must have made who have considered the condition of mankind in their various stages of advancement, that an increase in the physical comforts, and above all, the improvement in the diet, are the first symptoms of an onward movement in civilization. It has always appeared to me, that it is in vain to expect any other condition than that of retrogression among people, such as are too abundant in Scotland and Ireland, where the clothing is so deficient as to leave the extremities of the body, more particularly among the female classes, the educators of the community, in a state of nudity, and where the food is confined in a great measure to the watery potato, or the dry and unpalatable oat-cake.*

Maize bread may be made of good quality by a smaller admixture of flour than is necessary in the instance of oatmeal. For this purpose, it should be reduced to a fine meal,—finer than is usual in America. It may then be mixed with one-third its weight of best flour, and fermented in the usual way. When thus prepared, the best maize bread is always dark colored, and cannot be made much lighter than coarse wheat bread. The shade, however, is somewhat different

---

* By custom, it becomes more agreeable, but at first it is usually nauseous, especially to one who is not a native of the country.

from that of wheat, as it inclines more to a yellow tint. We may be quite certain, however, when we see what is called maize bread possessed of a white color, that it contains much more than one-third its weight of wheat flour mixed with it.   Even when one-half its weight of wheat flour is added to it, the dark color, characteristic of maize, is retained.   In these cases, of course the price of the bread must be higher than when a smaller amount of maize is present.

The whitest bread, however, is made by an inter-mixture of barley meal and wheat flour. The smallest amount of wheat flour in this mixture, which I have found requisite to make a good loaf, was one-half, al-though the quantity of flour may be diminished accord-ing to the increase in the richness of wheat in albumin-ous matter; an observation which, of course, applies to the various kinds of bread to which allusion has already been made. The most successful of these varieties of bread is, perhaps, that which is made with equal quantities of peas-meal and flour, so far as re-spects the exterior aspect. The last, however, is pala-table, and the specimen is a good example of a whole-some, condensed vegetable diet, and would probably answer as a substitute for animal food where the func-tions of the stomach are not materially impaired.

Upon similar principles, excellent biscuits may be made, either for rapid consumption, or for preservation, at a more moderate expense than when they are entirely composed of wheat flour. When a biscuit is formed of Indian corn, without any intermixture of wheat, the color has a yellow tint, which, however, in a great measure, disappears when wheat flour is added in the proportion of one-third When destitute of the pres

ence of wheat, it is not so consistent, and is apt to
crack and break off short. Oat-meal and barley-meal
biscuits may be produced also by mixture with wheat
flour. They require, however, a somewhat larger proportion of the latter, as their particles seem even less
adapted of themselves to cohere than those of the Indian corn. An admixture of a variety of meals forms
a very palatable biscuit, as it possesses a sweeter taste,
even without the artificial addition of sugar, than wheat
flour alone. Such biscuits are calculated to keep for
a longer or shorter time, according to the firing to
which they are subjected. In the former case they ·
are well calculated to keep at sea.

Bread of such a description may be made either by
the usual process of fermentation, or by the action of
hydrochloric acid upon sesquicarbonate of soda. In
many respects the latter process deserves the preference, when we consider the chemical nature of the two
methods.

The vulgar idea which yields the palm of superiority
to the former, does not appear to be based on solid
data, and it seems desirable, that in a case of so much
importance in domestic economy, the arguments in
favor of such an opinion should be subjected to a careful
experimental examination. Judging *à priori*, it does
not seem evident that flour should become more wholesome by the destruction of one of its important elements, or that the vesicular condition engendered by
the evolution of carbonic acid from that source, should
at once convert dough (if it were unwholesome) into
wholesome bread.

When a piece of dough is taken in the hand, being
adhesive, and closely pressed together, it feels heavy,

and if swallowed in the raw condition, it would prove indigestible to the majority of individuals. This would occur from its compact nature, and from the absence of that disintegration of its particles which is the primary step in digestion. But, if the same dough were subjected to the elevated heat of a baker's oven, 450°, its relation to the digestive powers of the stomach would be changed, because the water to which it owed its tenacity would be expelled, and the only obstacle to its complete division and consequent subserviency to the solvent powers of the animal system would be removed. This view of the case is fully borne out by a reference to the form in which the flour of the various species of *cerealia* is employed as an article of food by different nations. By the peasantry of Scotland, barley-bread, oat-cakes, peas-bread, or a mixture of peas and barley-bread, and also potato-bread, mixed with flour, are all very generally employed in an unfermented form with an effect the reverse of injurious to health. With such an experience, under our daily observation, it seems almost unnecessary to remark, that the Jew does not labor under indigestion when he has substituted, during his passover, unleavened cakes, for his usual fermented bread; that biscuits are even employed when fermented bread is not considered sufficiently digestible for the sick; and that the inhabitants of the northern parts of India and of Affghanistan very generally make use of unfermented cakes, similar to what are called *scones* in Scotland. Such, then, being sufficient evidence in favor of the wholesomeness of unfermented bread, it becomes important to discover in what respect it differs from fermented bread. Bread-making being a chemical process, it is from chemistry alone that we can expect a solution of this question. In the production of fermented bread a certain quantity of flour, water, and yeast, are mixed together, and formed into a dough or paste, and are allowed to ferment for a certain time at the expense of the sugar of the flour. The mass is then exposed in an oven to an elevated temperature, which puts a period to the fermentation, expands the carbonic acid, resulting from the decomposed sugar and air contained in the bread, and expels the alcohol formed, and all the water capable of being removed by the heat employed. The result gained by this process may be considered to be merely the expansion of the particles of which the loaf is composed, so as to render the mass more readily divisible by the preparatory organs of di gestion. But as this object is gained at a sacrifice of the integrity of the flour, it becomes a matter of interest to ascertain the amount of loss sustained in the process. To determine this point, I had comparative experiments made upon a large scale with fermented and unfermented bread. The latter was raised by means of carbonic acid generated by chemical means in the dough. But to understand the circumstances, some preliminary explanation is necessary. Mr. Henry of Manchester, in the end of last century, suggested the idea of mixing dough with carbonate of soda and muriatic acid, so as to disengage carbonic acid in imitation of the usual effect of fermentation; but with this advantage, that the integrity of the flour was preserved, and that the elements of the common salt required as a seasoner of the bread was thus introduced, and the salt formed in the dough.

The result of my experiments upon the bread produced by the action of hydrochloric acid upon carbonate of soda, has been, that in a sack of flour there was a difference in favor of the unfermented bread to the amount of 30 lbs. 13 oz., or, in round numbers, a sack of flour would produce 107 loaves of unfermented bread, and only 100 loaves of fermented bread of the same weight. Hence it appears, that in the sack of flour by the common process of baking, 7 loaves, or $6\frac{1}{2}$ per cent. of the flour, are driven into the air and lost. An important question now arises from the consideration of the result of this experiment: Does the loss arise entirely from the decomposition of sugar, or is any other element of the flour attacked?

It appears from a mean of eight analyses of wheat flour from different parts of Europe by Vauquelin, that the quantity of sugar contained in flour amounts to 5·61 per cent. But it is obvious that, as the quantity lost by baking exceeded this amount by nearly one per cent., the loss cannot be accounted for by the removal merely of the ready-formed sugar of the flour. We must either ascribe this extra loss to the conversion of a portion of the gum of the flour into sugar and its decomposition by means of the ferment, which is highly probable, or we must attribute it to the action of the yeast upon another element of the flour; and if we admit that yeast is generated during the panary fermentation, then the conclusion would be inevitable, that another element of the flour, beside the sugar, or gum, has been affected. For Liebig has well illustrated the fact that when yeast is added to wort, ferment is formed from the gluten contained in it, at the same time that the sugar is decomposed into alcohol and carbonic acid. Now, in the panary fermentation, which is precisely similar to the fermentation of wort, we might naturally expect that the gluten of the flour would be attacked to reproduce yeast.

A wholesale and palatable bread may be produced by the employment of ammoniacal alum, and carbonate of ammonia, or soda as a substitute for yeast. In this process the alum is destroyed by the heat : the bread is vesicular and white, and rises, according to the judgment of the baker, as well as fermented bread. It is obvious that none of the ingredients added can affect the integrity of the constituents of the flour ; an occurrence which may possibly happen in the preparation of bread by the common process of fermentation, as has been shown even to the azotized principles of the flour. The disadvantage of such a deterioration is sufficiently evident, if we view these principles as the source of nutrition in flour.

A good method of making unfermented bread is to take of flour 4 pounds. Sesquicarbonate of soda, (supercarbonate of the shops,) 320 grains. Hydrochloric acid, (spirit of salt or muriatic acid of the shops,) $6\frac{1}{2}$ fluid drachms. Common salt 300 grains. Water, 35 ounces by measure. The soda is first mixed with the flour very intimately. The salt is dissolved in the water, and added to the acid. The whole being then rapidly mixed as in common baking. The bread may either be baked in tins or formed like cottage loaves, and should be kept from one to two hours in the oven. Should the bread prove yellow, it is a proof that the soda has been in excess, and indicates the propriety of adding a small additional portion of acid ; the acid varying somewhat in strength. The same process may be employed in raising the other mixture previously recommended.

APPENDIX

TABLE I.—*Relations of the Food to the Products of Two Cows.*

	Grass.	Entire Barley.	Entire Malt.	Crushed Barley.	Crushed Malt.	Barley and Molasses.	Barley and Linseed.	Beans.
	lbs.	lbs.	lbs.	lbs.	lbs.	lbs.	lbs.	lbs.
Dry milk of 2 cows	82·99	56·25	49·64	90·00	85·15	51·38	54·96	27·00
Dry dung do.	239·79	201·78	633·16	360·22	339·65	231·96	225·15	105·89
Dry hay	713·50	582·80	527·00	780·66	745·50	420·33	436·65	225·62
Molasses	-	-	-	-	-	54·00	-	-
Dry grain	-	86·02	106·35	260·90	324·72	162·98	213·51	107·28
Total dry food	473·71	668·82	633·35	1041·56	1070·22	637·31	680·16	332·90
Ratio, dry food to milk	100 to 11·60	100 to 8·41	100 to 7·08	100 to 8·64	100 to 7·05	100 to 8·06	100 to 8·45	100 to 8·13
——, molasses to do.	-	-	-	-	-	9·44	-	-
—— to butter	2·74	1·82	2·07	2·11	1·52	2·19	2·15	2·25
——, dry hay to milk	-	9·66	9·60	11·50	11·40	12·33	12·60	12·00
——, dry grain to do.	-	65·40	46·53	34·60	26·90	31·50	25·70	25·23
—— to butter	-	14·32	12·40	8·40	6·33	6·45	6·57	7·00
Oil in grain	-	2·07	1·53	5·76	4·58	3·92	6·68	2·59
Wax in hay	57·30	37·78	34·17	25·95	17·64	10·86	10·39	5·35
Total wax and oil in food	-	39·85	35·70	31·71	22·32	14·78	17·07	7·94
Wax in dung	6·30	5·37	4·97	13·76	12·97	8·85	8·60	4·04
Butter	16·70	12·23	13·125	22·00	20·60	14·00	14·00	7·50
Excess of wax and oil in food	34·30	22·25	17·605	4·05	11·25	8·07	5·53	3·60
Excess of wax in dung and butter	-	-	-	-	-	-	-	-
	14 days.	11 days.	10 days.	16 days.	16 days.	10 days.	10 days.	5 days.

TABLE II.

Amount of Oil and Wax in the Food, and of the Butter.

BROWN COW.

Experiments.	Wax.	Oil.	Total.	Butter.	Ratio of Butter to Oil and Wax.
	lbs.	lbs.	lbs.	lbs.	
I. Grass - - -	28·68	—	28·68	9·71	1 to 2·95
II. Barley and grass - -	18·89	1·035	19·925	6·09	1 to 3·27
III. Malt and grass - -	17·085	0·765	17·850	5·50	1 to 3·24
IV. Barley, hay, and grass -	13·096	3·110	16·106	10·00	1 to 1·61
V. Malt and hay - -	8·190	2·306	10·496	9·353	1 to 1·12
VI. Barley, molasses, and hay -	5·38	1·80	7·180	6·289	1 to 1·14
VII. Crushed barley and hay -	2·64	1·20	3·84	2·80	1 to 1·37
VIII. Barley, linseed, and hay -	5·35	3·21	8·56	6·128	1 to 1·39
IX. Bean meal, and hay -	2·96	1·144	4·104	3·170	1 to 1·29

In this Table and the following the Casein and Water have been deducted from the Butter.

TABLE II.—(Continued.)

Amount of Oil and Wax in the Food, and of the Butter.

WHITE COW.

	Experiments.	Wax.	Oil.	Total.	Butter.	Ratio of Butter to Oil and Wax.
		lbs.	lbs.	lbs.	lbs.	
I.	Grass - - -	28·68	—	28·68	7·00	1 to 4·09
II.	Barley and grass - -	18·89	1·035	19·925	4·425	1 to 4·502
III.	Malt and grass - -	17·085	0·765	17·850	5·825	1 to 3·06
IV.	Barley, hay, and grass -	12·85	3·11	15·91	9·100	1 to 1·75
V.	Malt and hay - -	9·45	2·306	11·756	8·33	1 to 1·41
VI	Barley, molasses, and hay -	5·48	1·80	7·28	5·59	1 to 1·30
VII.	Crushed barley and hay -	2·63	1·20	3·83	2·42	1 to 1·58
VIII.	Barley, linseed, and hay -	4·98	3·21	8·19	5·859	1 to 1·39
IX.	Bean meal, and hay -	2·39	1·144	3·534	3·23	1 to 1·09

TABLE III.

Amount of Oil and Wax in the Food, and of the Butter in both Cows.

	Experiments.	Wax in Food.	Oil in Food.	Total.	Butter.	Ratio of Butter to Oil and Wax.
		lbs.	lbs.	lbs.	lbs.	
I.	Grass - - - -	57·36	—	57·36	16·71	1 to 3·43
II.	Barley entire and grass	37·78	2·07	39·85	10·53	1 to 3·78
III.	Malt entire and grass -	34·17	1·53	34·70	11·32	1 to 3·05
IV.	Barley crushed, grass, and hay	25·94	6·22	32·16	19·10	1 to 1·68
V.	Malt and hay - -	17·64	4·61	22·25	17·68	1 to 1·25
VI.	Barley, molasses, and hay	10·86	3·60	14·46	11·88	1 to 1·21
VII.	Crushed barley and hay	5·27	2·40	7·68	5·22	1 to 1·47
VIII.	Barley, linseed, and hay	10·33	6·42	16·750	11·98	1 to 1·39
IX.	Bean meal and hay -	5·35	2·28	7·64	6·40	1 to 1·19
				232·85	110·82	

TABLE IV.

Ratios of Food, Milk, and Butter.

BROWN COW.

	Milk every five Days.	Barley.	Grass.	Hay.	Dry Hay.	Butter every five Days.	Grain to Milk.	Butter to Grain.
	lbs.	lbs.	lbs.	lbs.	lbs.	lbs.	100 to	100 to
Barley crushed :								
1st five days -	115·68	42 3 malt	240	65	134	3·625		
2d do. -	105	45	26	153	136	3·33	255	
3d do. -	110·5	45	–	139¾	117½	3·935	245	
4th do. -	95·76	60		132¼	111	3·26	159	
	311·26	150		425	364·3	10·525	214	1428
Malt :								
1st five days -	97	42 3 barley		135¾	114	3·44	215	
2d do. -	96	54		129¼	108·5	3·60	177	
3d do. -	98·19	60		119	99·9	3·25	163	
	291·19	156		384	322·46	10·29	185	1514
Barley & molasses :		Barley.	Molasses.					
1st five days -	105·18	45	12	133	111·75	3·63	184	
2d do. -	98·5	45	15	137	115·00	3·63	164	
	203·68	90	27	270	226·75	7·26	174	1611
Barley & linseed :			Linseed.					
1st five days -	101·18	45	15	129	108·36	3·089	167	
2d do. -	104·00	35	25	136	114·25	3·228	173	
	205·18	80	40	265	222·61	6·917	170	1736
Bean meal :		Beans.						
1st five days -	99·72	56	4	148	124·32	3·69	166	1628

Note.—This and the opposite Table are read as follows:—During the second five days of experiment the Cow afforded 105 lbs. of milk and 3·33 lbs. butter, and consumed during that period 45 lbs. barley, 26 lbs. grass, 153 lbs. hay. The ratio of the barley to the milk is as 100 to 255, while the relation of the butter to the barley during fifteen days is as 100 to 1428, or 100 lbs. of grain would produce 225 lbs. of milk, and 1428 lbs. of grain would produce 100 lbs. of butter.

TABLE IV.
Ratios of Food, Milk, and Butter.

	Milk every five Days.	Barley.	Grass.	Hay.	Dry Hay.	Butter every five Days.	Grain to Milk.	Butter to Grain.
WHITE COW.								
	lbs.	lbs.	lbs.	lbs.	lbs.	lbs.	100 to	100 to
Barley crushed:								
1st five days -	109·68	42	240	65	134	3·19	272	
2d do. -	109·33	45	26	153	128·5	3·333	242	
3d do. -	110·68	45	–	172·5	144·9	3·376	246	
4th do. -	107	56		131·75	110·67	2·843	191	
	327·01	146	26	457·25	384·07	8·552	224	1538
Malt:		Malt.						
1st five days -	106·5	42		150	126	3·126	240	
		3 barley						
2d do. -	107·5	54		147	123·48	3·072	198	
3d do. -	111·5	60		147·5	123·9	2·937	185	
	325·5	156	3 barley	444·5	373·38	9·135	209	1715
Barley & molasses:		Barley.	Molasses.					
1st five days -	112	45	12	131·75	110·67	3·26	196	
2d do. -	112·5	45	15	142·25	119·49	3·26	189	
	224·5	90	27	274·00	230·16	6·52	192	1800
Barley & linseed:			Linseed.					
1st five days -	113 ·	45	15	117·5	98·7	3·406	188	
2d do. -	117·68	35	25	131·75	110·67	3·421	196	
	230·68	80	40	249·25	209·37	6·827	192	1760
Bean meal - -	115·628	56	4	146	122·64	3·76	193	1590

TABLE V.
Amount of Wax and Oil in different Kinds of Food, and in Dung.

	Wax per cent.		Oil per cent
Rye-grass - -	2·01	Barley - -	2·18
Rye-grass hay - -	2·00	Malt -	1·37
Moist grass dung -	0·312	Bean meal - - -	2·035
Moist hay dung -	0·600	Linseed meal - -	4·00
Dry grass dung -	2·67		
Dry hay dung - -	3·82		

TABLE VI.

Comparison between the Wax of the Food and the Butter, and the Wax in Dung.

Food.	Wax and Oil in Food.	Butter.	Wax in Dung.	Excess of Wax in Dung and Butter.	Excess of Wax in Food.
	lbs.	lbs.	lbs.	lbs.	lbs.
I. Grass - - - -	·57·36	16·71	6·30	...	34·35
II. Entire barley and grass -	39·85	10·53	5·37	...	23·95
III. Entire malt and grass -	34·70	11·32	4·97	...	18·41
IV. Crushed barley, grass, and hay	32·16	19·10	13·76	0·70	
V. Crushed malt and hay -	22·25	17·68	12·97	8·40	
VI. Barley, molasses, and hay -	14·46	11·88	8·85	6·27	
VII. Crushed barley and hay -	7·68	5·22	4·27	1·81	
VIII. Barley, linseed, and hay -	16·75	11·98	8·61	3·84	
IX. Bean meal and hay -	7·64	6·40	4·04	2·80	
	100·94	72·26	52·50	23·82	

From this Table it appears that when grass was employed as food there was much more wax in the food than wax excreted, and butter in the milk ; but as soon as hay was substituted for the grass, the butter and the wax in the dung together exceeded the amount of oil in the grain, and wax in the hay.

www.ingramcontent.com/pod-product-compliance
Lightning Source LLC
Chambersburg PA
CBHW021808190326
41518CB00007B/499